An Appraisal of the Skulls and Dentition of Ancient Egyptians, Highlighting the Pathology and Speculating on the Influence of Diet and Environment

Judith Miller

BAR International Series 1794
2008

Published in 2016 by
BAR Publishing, Oxford

BAR International Series 1794

An Appraisal of the Skulls and Dentition of Ancient Egyptians, Highlighting the Pathology and Speculating on the Influence of Diet and Environment

ISBN 978 1 4073 0282 9

BAR Publishing is the trading name of British Archaeological Reports (Oxford) Ltd. British Archaeological Reports was first incorporated in 1974 to publish the BAR Series, International and British. In 1992 Hadrian Books Ltd became part of the BAR group. This volume was originally published by Archaeopress in conjunction with British Archaeological Reports (Oxford) Ltd / Hadrian Books Ltd, the Series principal publisher, in 2008. This present volume is published by BAR Publishing, 2016.

Printed in England

BAR
PUBLISHING

BAR titles are available from:

BAR Publishing
122 Banbury Rd, Oxford, OX2 7BP, UK
EMAIL info@barpublishing.com
PHONE +44 (0)1865 310431
FAX +44 (0)1865 316916
www.barpublishing.com

I wish to dedicate this book to my husband, Victor, for innumerable reasons,
but mainly for his love, encouragement and support.

Contents

CHAPTER 5. MATERIALS AND METHODS

CHAPTER 6. RESULTS

CHAPTER 7. DISCUSSION

CHAPTER 8 CONCLUSIONS

List of Tables

Pages 71-89

List of Figures Pages 90-135

Preface

When I first studied Ancient Egyptian History, I was struck by the fact that, with the exception of a blind harpist or an occasional adipose figure, the Ancient Egyptian, was portrayed as healthy and fit with a superb physique. However, the reality was somewhat different. It has been discovered in previous studies of the mummies and the profusion of skeletal material which are available in many collections that their lives were far from ideal and many died in pain with diseases found in modern man. Then there are the many medical papyri which give prescriptions for treatments. Some were magical and were, in reality, spells to rid the sick individual of possession by a malign spirit. However some were rational and were passed on from doctor to doctor.

As a dental surgeon I decided to look at the dental health in the population which covered a period of 4000 years.

This research in the present publication which originates in my doctoral thesis presented to the University of Manchester in 2002 was undertaken to investigate whether changes in the diet over a period of 4000 years had a direct effect on the dentition of the ancient Egyptian. The abundance of specimens in various collections made it possible to examine complete skulls to detect dental and bone pathology which may have been influenced by disease and the environment in which they lived.

Analysis of the literature of past surveys carried out in tombs has revealed much information. Tomb paintings symbolised an ideal presentation of food for the afterlife. There are scenes illustrating agriculture and irrigation of the land. In museum collections there are papyri listing rations allotted to workmen and soldiers.

Of particular importance are the burial goods. There are flagons containing dried remnants of wine and beer. There are mummified joints of beef. Offerings of fruit and grain are identifiable and have been analysed. Bread offerings, found in abundance, have been investigated to differentiate organic and inorganic components. From archaeological excavations, butchered bones from a variety of animals have been identified and the burial sites give clues as to the extent of the fertile area of the Nile Valley at different periods during the millennia.
This historical evidence has been examined to evaluate the extent of medical knowledge at various periods and this has been related to the pathology found.

This study has been augmented since its completion by further information that has been discovered about those with the title of dentist. Also included is an investigation into the aetiology of the dislocation of the mandible in the Ancient Egyptian, the treatment of which is discussed in the Edwin Smith Papyrus.

Acknowledgements

First of all, I acknowledge the friendship, inspiration and encouragement of Professor Rosalie David who introduced me to Egyptology and supervised my research into the ancient Egyptian history.

For authorising access to the Egyptological specimens in the Duckworth Collection in Cambridge, thanks are due to Dr. Robert Foley, the Director. Maggie Belatti, technician to the Palaeoanthroplogical Laboratory, gave invaluable help and guidance with lists and the locations of the specimens.

Thanks are also due to Dr. Robert Kruszinski and Theya Molleson for allowing me to examine the Graeco-Roman collection in the Natural History Museum and for their interest in the project.

Dr. Vic Garner kindly facilitated the analysis of the hair sample in the Mass Spectrometer at the Hall Analytical Laboratory and proof read the section about the technique.

Dr. Tatiana MacFarlane from the Department of Oral Health and Development at the University Dental Hospital was infinitely patient with her explanations and help with the statistics.

Finally, and significantly, I owe a huge debt of gratitude to Professor Fraser McCord, for his supervision and reassurance in my many moments of doubt. Without his guidance and unstinting help this thesis would never have been completed.

Judith Miller PhD MSc BDS retired from dental practice in 2002 and is now an Associate at the Wellcome Trust Centre for the History of Medicine and a member of the Manchester University Mummy Research Team. She teaches dental studies to the students on the Biomedical and Forensic Egyptology MSc course.

Introduction

The aim of this study was to investigate the influence of diet on the skull pathology and dentition of the Ancient Egyptian and detect significant differences in dental disease in successive time periods. A historical and quantitative assessment of this information has been made.

The dentition is the only part of the skeleton which, in health, is exposed to the external environment. Diet has been shown to affect the skeleton and dentition either directly or indirectly. Indirect effects are seen in dietary deficiencies which may result in scurvy, rickets, osteomalacia or lines of arrested development in the developing skeleton and in hypoplasia and hypocalcification in the developing teeth. Direct effects of diet on the dentition can cause caries, wear and periodontal disease.

Of the mineralised animal tissue, teeth are more resistant to post-mortem decomposition than bone and appear to be unaffected by chemicals used in mummification. When all other structures of the body have disintegrated, the dentition is often all that remains intact on excavation. From these can be gleaned much information about the diet and health of a population at a specific time period in antiquity. In addition to this, cultural information can be acquired from tomb models, paintings and stelae and the dietary insight from the food offerings found, often in an excellent state of preservation, in the tombs. These sources have been utilized in the research for this thesis which has examined the dental condition of skulls representing the Ancient Egyptians over a period of almost 5000 years.

For this appraisal, a selection of the Egyptian skulls from the Duckworth Osteological Collection, housed in the School of Biological Anthropology in Cambridge, were examined. These samples were augmented by skulls from the Department of Palaeopathology at the British Museum (Natural History). These collections between them contain over 5000 skulls from sites in Upper and Lower Egypt dating from the Predynastic to Ptolemaic period, a time span of almost 5000 years. Specimens from the Predynastic, Old Kingdom, Middle Kingdom, New Kingdom and Graeco-Roman Dynasties were considered as separate groups. When possible, more than one site from each period was investigated. The findings were related to known differences in diet in the various geographical regions and time periods. An example of this was the transition from Neolithic hunter-gatherers to settled agriculturists and pastoralists around 4000 BC. The cultivation of grain as a main source of food and the concomitant reduction of hunting for animals resulted in a decrease in the protein and an increase in the carbohydrate component in the diet.
In this thesis, reference is made to the effects of the variations in diet engendered by this transition together with the periods of drought and flooding which have been recorded throughout the chronicled history of Egypt. There were years of abundance when the inundation of the Nile was optimum. Interspersed between years of plenty were periods of famine when there was harvest failure caused by drought and flooding. The diet over the millennia varied and this may have caused a detrimental affect on the dentition. Other modifications arose within the Dynastic Period resulting from the changes in the culture following the invasion of peoples from neighbouring countries, the capture of prisoners and expanding trade with the addition of imported crops.

Direct influences on the dentition were the contamination of the food by wind-blown sand, the methods of food preparation and the constituents of the diet. The results of the assessment indicate that dental disease was significantly more severe than the living conditions indicated and reasons for this are examined and discussed.

Table 1 shows the different time periods with the dates, the dynasties and the sites included in the survey. Those selected were times of stability and prosperity in Egypt. The Intermediate Periods were excluded because there was general turbulence with ineffectual rulers and foreign intruders. This resulted in an unstable population because there was an influx of peoples from other countries.

Table 1: ***Time Periods and Sites***

PERIOD	DATE	DYNASTY	SITE	NO OF SKULLS
Predynastic	*5000-3100BC*		*Badarian* *Naqada* *Hierakonpolis* *Nubia*	*40* *40* *03* *08*
Archaic ***Old Kingdom***	*3100-2686 BC* *2686- 2613 BC*	*1,11* *111-V1*	*Abydos* *Qau* *Nubian (Pearson)* *Tarkhan* *Dishasha/Aswan*	*30* *20* *30* *30* *10*
Middle Kingdom	*1991-1786 BC*	*X11*	*Kerma, Sudan* *Qurna el Tarif*	*70* *20*
New Kingdom	*1567-1085 BC*	*XV111-XX*	*Qurna* *Misc.* *Memphis*	*50* *5* *34*
Late Period	*664-332 BC*	*XXV1-XXX1*	*Gizeh*	*60*
Ptolemaic Period ***Roman Period***	*332-30BC* *30 BC-400 AD*		*Misc.* *Hawara.*	*17* *58*

To begin the historical review, previous translations from the Hieratic of the medical papyri which contained dental references were examined

Chapter 1

The Medical Papyri

'A practitioner of medicine is a physician. His procedures depend on both experience and theory' (Sigerist, 1951). Nowadays a physician can practise medicine only if he is authorised which means that he must have a qualification and be fully licensed. The history of the evolution of medicine in Egypt demonstrates how the experience and theory developed.

1.1 The development of medicine

The earliest Egyptian medicine was composed of magic, religious and empirical elements. However, the different components were not separated. The magician used drugs and the physician reverted to magic and prayer. Breasted (1931) wrote about the Code of Hammurabi which affirms the existence of a practice of medicine in Babylon probably dating from about 2000 BC. No trace of this original document has been found and only incomplete fragments of much later copies of Babylonian medical texts from the 7th Century BC are in existence. Therefore, the earliest medical documents which survive are the ancient Egyptian medical papyri. The first of these dates from the 18th Century BC. Some of these are medico-magical. Diseases were personified and addressed by the magician. The body was either possessed or poisoned by a demon. Dawson (1967) said that *'magic was used to expel the intruder'*. Gardiner (1961) described the two components used as the *'oral rite'* and *'manual rite'*. The oral rite was a spoken text, the spell, in which the demon was harangued. The manual rite warded off evil by using a charmed artefact, such as an amulet, which was laid on the body of the patient or by a substance, which was given to the patient to swallow or to apply to a wound. These rites were often used in combination. The prescriptions were made unpleasant deliberately in order to make the substance hostile to the demon which occupied the body. The magician had other options if he did not, at first, succeed. Some of these prescriptions, however, were found to be effective. The oral rite gradually became less important and thus began the transition from magician to physician, the ***swnw***. Before the advent of modern medicine, *'despite the hazardous remedies, patients often became better because, modern experts say, the authority of the healer and the ritual of treatment affected the balance of mind and body for the better. With notable exceptions, such as opium for pain, or quinine for malaria, or smallpox injections, remedies owed their success, according to historians of medicine, almost entirely to placebo effect'* (Cornwell, 2000).

1.2 The medical papyri

Only ten medical papyri have, as yet, been discovered and all are of unknown authorship. Most of the papyri may have come from the tombs of doctors but have no association with identified doctors. Nunn (1996) suggested that the possible explanation for their rarity is that some may have been destroyed during early tomb robberies and many others may have been destroyed in the fires in the library in Alexandria in 47 BC and 389 AD. Ghalioughui (1980) has dated the papyri from the 11th to the 19th Dynasties (2040 BC to 1200 BC). Some are mainly magical, some medical and some a mixture of both. The *Turin Papyrus* from the 11th Dynasty is a magical formulary about eye diseases, snake bites and ingested fish bones. The *Kahun Papyrus*, from the 12th to 13th Dynasty, is the only other extant text which is dated earlier than the New Kingdom. However, there are indications that all the papyri were based on much earlier medical texts, the original manuscripts having being lost, so that those extant are copies. They were edited and new prescriptions were added and unsuccessful ones were omitted (Sigerist, 1951). The Ebers and Berlin Papyri mention the *'Book on the Vessels of the Heart'* (Ghalioungui, 1980) which is said to have been found under the feet of Anubis at Letopolis and brought to his Majesty, the King of Upper and Lower Egypt, Usephais from the 1st Dynasty. Murray (1972) recorded that Manetho, an Egyptian priest and historian, had written a history of Egypt for Ptolemy Philadelphus (c. 270 BC) and ascribed the medical texts to Athotis, son of Menes, the first pharaoh of the 1st Dynasty. That there were earlier medical texts than the papyri is confirmed by the archaic terms which are used and which are explained in several texts by the *'glosses'* which are commentaries on those ancient terms.

1.3 Dental references in the medical papyri

Weeks (1980) stated that five of the ten known medical papyri contain dental references. These are the *Ebers*, the *Edwin Smith*, the *Kahun*, the *Hearst* and the *Berlin* Papyri. Two of these, the *Ebers* and the *Edwin Smith* Papyri, are thought to have a common source as they both refer to *'The Secret Book of the Physician'*. Similar terms, which are archaic compared with the rest of the text, have been used in both documents. Their origin may have been an earlier treatise some 1000 years older and there is unfounded speculation (Weinberger, 1947) that it was the work of Imhotep who was a vizier and architect at the time of Djoser (2667 BC). The design and construction of the 1st pyramid, the Step Pyramid, was attributed to him. He was also a noted physician, deified as a god of medicine in the Late Period (664 –332 BC).

During the Graeco-Roman period he was also identified with the Greek god of medicine, Asklepios (David and David, 1992). Greek medicine encompassed much of the earlier Egyptian medicine and was a progression from that period. It therefore follows that the foundation of modern medicine had an even earlier origin than has been supposed.

The medical papyri differ in both quality and content. Most are written in Hieratic, which is a cursive form of Hieroglyphs, used in legal and religious documents. The *Edwin Smith Papyrus* is a factual document with 48 described surgical cases. These start with injuries to the head and continue with injuries to the thorax and abdomen. It is, in essence, a treatise of surgical conditions which are logically discussed. On the verso there are spells and incantations. The *Ebers Papyrus*, which is the most complete, is in excellent condition. It is a handbook for physicians which describes medical conditions in a random order, and in addition, there is a compilation of extracts, recipes and notes from some forty different sources. However, Dawson (1967) considered that it is not a treatise as had been previously claimed but is a miscellany of disparate entities. Several of the same conditions are described in both the *Edwin Smith* and *Kahun Papyri*. The *Kahun Papyrus* discusses, in Hieratic, mainly gynaecological conditions but there is also a text on veterinary topics written in Hieroglyphs, which is usually reserved for religious documents. The papyrus is in poor condition with many missing fragments. The *Hearst Papyrus* has almost half of its contents in common with the *Ebers Papyrus* but includes some spells and incantations. The *Berlin Papyrus* was written by a scribe who was also a doctor. It has parts allied to the *Ebers* and *Kahun Papyri* (Nunn, 1996). It is not arranged in any specific order. The other medical papyri are the *London*, the *Carlsberg*, the *Chester Beatty* and the *Ramesseum Papyri* but these are not relevant to this dissertation, as they do not contain specific dental remedies. The *Chester Beatty Papyrus*, however, which has medical prescriptions which are mainly magical, contains special incantations for headaches. One is for ***ges-tep*** which means 'half head' from which the word 'migraine' is derived. As one cause of migraine is considered by some authorities to be related to dental occlusion it is worthy of a mention (Gray *et al.*, 1994).

There are also papyri from Egypt from the 2nd Century AD which are written in Greek, one of which has a prescription for a tooth powder and another which has a treatment for dislocation of the mandible.

1.4 Vocabulary used for dental terms

In the medical papyri discussed above, there are words for teeth and the supporting tissues which were frequently used. Translations of these terms are given below.

Table 2a: ***Some common terms used in the medical papyri. (Weeks, 1980)***

WORD	*MEANING*
ibḥ	The word customarily used for tooth
nḥdt	Translated as a molar. The correct translation is thought to be canine as it is also used to refer to an elephant's tusk, which is prominent. (Tusks are now known to be incisors.)
ibḥy	'One who is concerned with teeth'.
Iry ibḥ	'One who treats teeth'. The difference between these terms is unknown.
w3b	Root of a tooth and is also the root of a plant.
ḥ'w	Gums or flesh.
3rt	Translated as mandible
'rty	Has been translated as the two sides of the mandible but is thought to be the maxilla and the mandible.
Wgyt	Meaning 'the bones of the jaw' but may include the surrounding soft tissue.
3m't	Forked bone such as the ramus. This word is derived from ***3m'-***
3m'-bird	An unidentified two-toed bird thought to be related to the ostrich. It also refers to other bones such as the scapula and the ulna, which have forked joints.
Ns	Tongue. (The most commonly used term.)
šnk & ***šsr***	Tongue, as used at a later period. It is not known if there are differences in their meanings.
šs3w	Tongue (in the Pyramid Texts.)
r	The mouth
ḥngg	Has been translated as the buccal cavity but is more likely to be the hard palate.
špty	The lips
in'	The chin
mwyt r3.	Saliva
mwyt	Urine!

The ancient Egyptian physician realised the importance of air and water for survival. He thought that air travelled directly into the heart and entered the blood stream. Various afferent channels then carried it to the organs. These channels were called ***metu*** and a set of efferent ***metu*** carried the waste products from these organs to the surface. The ***swnw***, or physician, did not distinguish arteries from veins or nerves. The waste products were the ***wḫdw*** which were of great concern to the ancient Egyptian. In the *Ebers Papyrus* there is a chapter about *'expelling the wḫdw.* The physician thought that the *wḫdw* were putrefying faeces, which would enter the *metu* and infect the heart and organs, thus causing disease. One of its manifestations was considered to be dental disease. The term 'Blood-Eater' appears in the medical texts and was the effect of ***wḫdw***, or pus, on the blood, resulting in the clotting of blood. In chapter 89 of the *Ebers Papyrus* there is a prescription to treat 'blood-eating in a tooth'. This clotting was understood to be pathological and not part of the healing process. The physician tried to prevent the formation of a scab so that the ***wḫdw*** could escape.

The following dental prescriptions are from the papyri in the order from the earliest to the latest.

1.5 The medical papyri with dental references

1.5.1 The Kahun Papyrus

The *Kahun Papyrus*, the oldest medical papyrus with dental references dating from the 12th to the 13th Dynasty, was written in approximately 1950 BC. It was found in the Fayum in 1898 and was reported to be in very poor condition. There are two cases which mention teeth (Stevens, 1975). A new translation by Stephen Quirke has not yet been published.

The words in capitals were written in red ink in the papyrus.

Case no. 5 is entitled,

'*INSTRUCTIONS for a woman suffering in her teeth and her gums and she cannot open her mouth. YOU SHOULD DECLARE ABOUT HER: 'This is acute pains of the womb.' YOU SHOULD PRESCRIBE FOR IT: --------(*this information has been lost)-----*then fumigate her over oil and frankincense in a dᶜdᶜ bowl, and pour into her vagina the urine of an ass who has created his fellow the same day as he passed it. Etc.*' Manniche (1993) translated a ***dᶜdᶜ***- bowl used for fumigation as, possibly, a double pot. In the lower pot a heated stone was placed and the oil and frankincense were poured over it. The second pot was fitted inside and had a hole in its base in which a straw was inserted for the fumigation.

The translation by Grapow, von Deines and Westendorf in 1958 is rather different (Weeks, 1980). Frankincense is used in the first translation and terebinth (turpentine) in the second. Also it seems that the urine in the first prescription is fresh after the delivery of a newborn ass and no mention of this is made in the second prescription. In Grapow *et al.,* (1958) ***ibḫ***-teeth and ***nḥḏt***-teeth are mentioned but there is no reference to 'gums'. These terms will be reviewed below. As the patient is referred to in a gynaecological document, it is likely that she was of childbearing age. This is the period in which a third molar might be erupting. An explanation for her inability to open her mouth might be pericoronitis, surrounding the crown of a partially erupted tooth, causing oedema and resulting in trismus. Another cause of trismus or 'lock jaw' might be tetanus. This is a neuromuscular disorder and it may follow trauma and is caused by an exotoxin of Clostridium tetani penetrating an open wound. Spasm may occur in the muscles near the injury (Cecil & Loeb, 1959[1].

Case no. 33 is

'TO PREVENT TOOTHACHE IN A WOMAN WHO HAS CONCEIVED: *beans ground up with ------ for her teeth until the days of giving birth. This is a cure for toothache that has been really successful countless times.'* It is possible that the missing word could be ***tjeret***, willow, which is the precursor of aspirin.

1.5.2 The Edwin Smith Papyrus

Breasted wrote in 1930 that the *Edwin Smith Papyrus* dates from 1700 BC. Edwin Smith bought it in 1862 from a dealer in Luxor. It is likely that it had been found in a tomb in Thebes. After his death, Smith's daughter gave the manuscript to the New York Historical Society where Breasted translated it. He also produced a facsimile and commentary. Sigerist (1951) thought that this is one of the most adept editions of an ancient text. The papyrus is 4.68 metres in length and is 30 centimetres high. There are 17 columns of writing on the recto and $4^1/_2$ columns on the verso. The recto is a surgical treatise of cases rather than recipes. The cases are arranged in order from injuries of the head down through the body. The cases described are all injuries which have occurred in men, and it has been suggested that the author may have been an army doctor. The treatment is rational and is laid out systematically. Each starts with the title and begins, *'Instructions concerning----'* and is followed by the examination which states that *'If thou examinest a man having --- thou shouldst---'* and this would be followed with how the examination would be performed. Then the diagnosis is made. *'Thou shouldst say concerning him. One having ---- '.* There are three possible verdicts given. *'An ailment which I will treat'* or, *'An ailment with which I will contend',* or, *'An ailment not to be treated'.* The treatment is then prescribed and at the end of the treatment section there are three possible periods of observation. *'Until he recovers.'* – *'Until the period of his injury passes by.'* –

and '*Until thou knowest that he has reached a decisive point.*'

There are 69 *glosses* which form a dictionary to explain terms which were archaic when the papyrus was written in the New Kingdom. The content of the papyrus was probably compiled in the Middle Kingdom, many centuries before, and the original treatise on which that was thought to have been based was already old. The *Ebers Papyrus* is the only other papyrus to have glosses. It has 26 from the same original treatise. They are introduced, '*with regard to*'.
Two cases are of interest to the dental surgeon.

Case 24

Title
Instructions concerning a fracture in his mandible

Examination
If thou examinest a man having a fracture in his mandible, thou shouldst place thy hand upon it. Shouldst thou find that fracture crepitating under thy fingers.

Diagnosis
Thou shouldst say concerning him: 'One having a fracture in his mandible, over which a wound has been inflicted, ------- and he has a fever from it. An ailment not to be treated.'

Breasted has suggested that the words missing may describe the injury in more detail and, because of the fever, no recovery was likely.

It may have been a compound fracture and, because of the introduction of secondary infection, the patient would have died as the result of septicaemia.

Case 25

Title
Instructions concerning a dislocation in his mandible

Examination
If thou examinest a man who has a dislocation of his mandible, shouldst thou find his mouth open and his mouth cannot close for him, thou shouldst put thy thumbs upon the ends of the two rami of the mandible in the inside of his mouth, thy two claws under his chin, thou shouldst cause them to fall back so that they rest in their places.

Diagnosis
Thou shouldst say concerning him: 'One having a dislocation in his mandible. An ailment which I will treat.

Treatment
*Thou shouldst bind it with **ymrw**, and honey every day until he recovers.*

The reduction of the dislocation is by the only procedure possible and so it is still used today. The technique is also described by Apollonius of Kitium in the 1st Century B.C. in his '*Illustrated Commentary on Hippocrates*'. The drugs used will be discussed later.

On the verso of the papyrus are spells and recipes. These are similar to those in the other medical papyri which do not include an examination of the patient but contain only the diagnosis and the treatment.

1.5.3 The Ebers Papyrus

The *Ebers Papyrus* was written about 1550 B.C. This is considered to be the best preserved, most complete and largest of the papyri. It is 20 metres in length and 30 centimetres high and is arranged in 108 columns of 21 lines. It was purchased in Luxor by Edwin Smith in 1862 and it is possible that it had been found in the same grave as the *Edwin Smith Papyrus*. He sold it to Professor Georg Ebers who named it after himself. He did not translate it but wrote an introduction to it and published a facsimile edition. Joachim translated it in 1890, but because the knowledge of Hieratic at this time was not developed, the interpretation is not accurate and presumptions have been made. In 1913 Wreszinski translated the papyrus into Hieroglyphs. The translation by Ebbell was published in 1937. In 1987 a more accurate translation was made by Ghalioungui but this, unfortunately, has not been published. Dr. Benson Harer, to whom Ghalioungui's widow gave one of the only two printed copies of his manuscript, has generously provided a photocopy of the dental prescriptions for this dissertation. Sigerist (1951) has suggested that the Ebers Papyrus is proof that not all physicians were specialists as stated by Herodotus (*vide infra 3.1*).

Ebbell divided the papyrus into several disparate sections such as treatment for internal diseases, prescriptions for eye diseases, skin diseases and diseases of the extremities. There are three grades of practitioner in the papyrus. The ***swnw*** was recognised as a physician. The ***Sekhmet priest*** was most likely to be a surgeon and the term is used in the *Edwin Smith Papyrus*. The ***s3w*** was the magician and used spells and amulets. The compiler of the papyrus also opened sections with '*the beginning of the physician's secret*' or '*the beginning of remedies prepared for women*', and was probably copying earlier medical books.

Prescriptions in the Ebers Papyrus

The translation by Ebbell is quoted.

In column 86 of the papyrus it states '*the beginning of remedies to expel illness of the tongue: milk, rinse the mouth and spit out.*' There is no list of symptoms. The diagnosis is given although it is difficult to imagine what the problem was. The milk could have been used to sooth the tongue if it had been burnt. Another remedy for the tongue was *'frankincense, cumin, yellow ochre, grease of goose, honey, water, rinse the mouth for 9 (days).'*

Many of the explanations of the drugs come from the glossary by Estes (1993), which is based on several sources.

Frankincense, a fragrant resin, which was used in the treatment of the tongue, was used non-selectively for problems with ***metu*** (vessels) and, because of its fragrance, may have been used if there was a smell from the ***wḫdw*** (waste products).

Yellow ochre is an iron oxide which was used to amend problems of the ***metu.***
Goose grease was a vehicle for active drugs.

Honey has been found to have an antibacterial effect as it is hypertonic and dehydrates the bacteria by osmosis. Honey would therefore also have reduced inflammation in swollen, infected gingivae. It contains propolis from the nectar which was used as an antiseptic after surgery until the end of the 19th Century. Other constituents of honey with antibacterial properties are gluconic acid and hydrogen peroxide. It has been shown that one part of honey to two parts of animal fat can arrest the growth of Staph. Aureus and E. coli (Estes, 1993).

Column 89 has eleven dental remedies. There are two prescriptions to *'fasten a tooth':*

1. *'Powder of ammi, yellow ochre, honey, are mixed together and the tooth is filled (?) therewith'* and *'scrapings of millstone, yellow ochre, honey, the tooth is filled therewith.'*

Ammi is an aromatic herb but Ghalioungui translated the word as emmer seed, which is wheat.

Millstone scrapings were used for superficial injuries. Yellow ochre was used as above. These substances would make a paste which would probably set and act like a splint.

The second prescription is

2. *'To fasten a tooth; frankincense, yellow ochre, malachite are pounded and applied to the tooth.'*

Ghalioungui (1987) translated frankincense as Terebinth-resin.

There is a prescription to 'expel eating ulcer on the gums and make the flesh grow; cow's milk, fresh dates, manna, it remains during the night in the dew, rinse the mouth for 9 days'.

Manna was the translation made by Ebbell but Estes thought that carob was the intended meaning. Carob was used for gut disorders. Ghalioungui interpreted the instructions as *to be chewed and spat out.* This is a description of a masticatory (*vide* 2.5.5).

He also translated the problem as *ulcers in the teeth.* This could be the description of the exposed pulp of teeth after the attrition of the secondary dentine. The destruction of the alveolar bone and 19aemorrh by a periapical abscess which had pointed might be the reason for the necessity for *the flesh to grow.*

Dates were used for gut disorders but not for skin wounds.

Another prescription is '*to strengthen the gums and treat the gums: celery/, dwꜣt/, sweet beer /, chewed and spit out.'* Ghalioungui translated it as 'to consolidate the teeth' and explained that there was no specific name for gums but that they could be named either ***ibḥ*** (tooth) or ***ḥꜥw*** (flesh).

Celery, as translated by Ebbell, was thought by Estes to mean calotropis which is giant milkweed. This is an emetic and a heart stimulant.

Beer was made from barley, wheat or dates. It may have been used to dissolve or dilute another substance, but it was also a diuretic.

ḏꜥrt has been translated by Ebbell and Ghalioungui as colocynth, the fruit of citrullus colocynthis which is a strong laxative.

'Another remedy to treat 'blood-eating' in the tooth: ḳbw, ḏꜥrt, gum, fruit of sycamore, inšt, water, it remains during the night in the dew, rinse the mouth for 4 days.' Ebbell described *'Blood-eating'* as scurvy. This could have referred to bleeding gums, as it was not always clear whether tooth or gums was the correct translation. One of the effects of scurvy, which is caused by a deficiency in ascorbic acid, is haemorrhagic aemorrh. Ghalioungui interpreted the method of treatment as chewing not rinsing.

Gum was a binding agent and it could have been used to form a masticatory. The fruit of sycamore was used for disordered *metu.* It is still used in Egypt for skin problems and burns.

inšt has been translated as aniseed (Manniche 1993). It was chewed as a remedy for toothache.

No translation has been found for ***ḳbw*** from the *Ebers Papyrus* or ***ymrw*** from the *Edwin Smith Papyrus* but it has been suggested that they are minerals.

1.5.4 The Hearst Papyrus

The Hearst Papyrus was found by a peasant who brought it to the Hearst Egyptian Expedition near Der-el-Ballas in

1901 to thank the expedition for allowing him to take ***sebaḫ***, fertilizer, from the waste dumps of the expedition. The papyrus had been found two years earlier in a pot near house walls when he had been previously searching for sebaḥ. George Reisner (1905), the leader of the expedition, named the papyrus, the *Hearst Papyrus*, to honour the mother of Randolph Hearst who had sponsored the work of the expedition (Nunn, 1996). The papyrus dates from the reign of Tuthmosis III in the 18th Dynasty, from 1490 BC. The outside of the papyrus was in poor condition, probably damaged when it was found, but fragments, which were broken off on the journey to the expedition, were unearthed from the head-cloth in which it was carried. Reisner was of the opinion that only one or two pages had been lost and that it had been compiled at a similar period to the *Ebers Papyrus*. There is one prescription of dental interest. This is column 1, line 7. This has been translated as '------*which has fallen downwards; mm one portion; ᶜᶜm one portion; gum one portion. Apply to the tooth.'* (Quirke, 1999). The *Ebers Papyrus* has similar prescriptions 'to consolidate' a tooth and Reisner translates it as a prescription for a tooth which has fallen out. It is likely to be a tooth which has become loose due to periodontal breakdown.

Mm is ammi which is translated as emmer seeds by Ghalioungui. I have not found a translation for ***ᶜᶜm.*** Gum is a viscous secretion from plants which becomes firm when it dries. It may have held the tooth rigid in a splint.

1.5.5 The Rylands Greek Papyrus

Dawson wrote about later medical documents, one of which is the *Greek Papyrus* in the John Rylands Library, Manchester (Hunt, 1911). Case no.29a is for a tooth powder. Unfortunately many words are missing from the fragment. It has been translated by Murphy (1998). '*For scabs in the mouth, the cup is scented with old wine. Having washed the mouth out of the very sweet------then having chewed up-------wash out the mouth. For stavesacre, white pepper------tooth powder, smooth Egyptian beans----of sandy---having made an incision----- For diseased uvulas, wheaten flower---smooth like granules---having rubbed in---Acacia, having made an incision—For thrush-----'*Acacia has astringent properties. Stavesacre (staphisagria- wild dry grapes) is usually known as larkspur which may have had an antiseptic effect. Egyptian beans were used for skin and musculo-skeletal disorders. (Estes, 1993).

1.6. Anatomical knowledge

The Egyptians had acquired a good knowledge of anatomy because, during the process of mummification, the viscera were removed and preserved by the embalmers although the physicians were not involved. The organs of the body were identified but no differentiation was made between blood vessels and nerve fibres and therefore the physiology was not understood. Also, observation of the organs occurred when treating traumatic injury. The Hieroglyphs representing the organs were often pictures of mammalian organs rather than human organs. Warren Dawson (1967) suggested that this was because the Egyptian knowledge of animal anatomy was older than their knowledge of human anatomy, since animals were prepared for food and sacrifice from an earlier date than mummification was practised. These symbols will be discussed below.

During the Ptolemaic Dynasty, Greek physicians came to Egypt to perform dissections which were not possible in Greece. In both Egypt and Greece defilement of the dead was prohibited. Egyptian and Greek medicine were rather disparate. When the Greeks settled in Alexandria during the Ptolemaic period (305-30BC) the Greek physicians treated the Greek ruling classes and the Egyptian physicians treated the Egyptian population. However the Greeks built a shrine at Deir el Bahri to Imhotep, the Egyptian god of healing, and identified him with Asklepios. During that period a medical school developed at Alexandria and this is where Galen studied in the 2nd Century AD. In the 3rd cent BC Herophilus, a follower of Hippocrates, came to Alexandria and studied both dissection and the vivisection of criminals given to him for this purpose by the Ptolemaic kings. He made many discoveries about anatomy and physiology, identifying seven of the cranial nerves and distinguishing the motor and sensory function of the nerves. He timed the pulse using a water clock. Although he did not understand the circulation of the blood, he noted the differences between veins and arteries (Nunn, 1996). His methods, although greatly increasing the contemporaneous knowledge, were somewhat lacking in compassion.

The medical papyri give some indication of medicaments that may have been in use for the recommended treatments. What is unknown is how available these prescriptions were and for what section of the population. The following chapter deals with what is known about the geography of Egypt during its predynastic and dynastic history, the crops grown and the medical conditions which can be attributed to climate and diet.

Chapter 2

Geography, Climate and Diet in Ancient Egypt and the Consequential Medical Conditions

2.1 Geography of Egypt

Herodotus (Selincourt, 1972) said that the land of Egypt was the *'gift of the river'* because the greater part of inhabitable Egypt had been built up from silt carried by the inundation of the Nile. Much of the country is desert, but the Nile Delta in Lower Egypt and the Nile valley are fertile. The contrast between the fertile land, the Black Land, *Kemet*, which was the original name for Egypt, and the desert, the Red Land, *Desheret*, is marked. The Nile is some six hundred miles long as it meanders from Aswan in Upper Egypt to Cairo. From Cairo to the Mediterranean through the Delta there is a further one hundred and fifty miles. The fertile ground on each side of the river can be cultivated, at its greatest width, for only thirteen miles across the valley. The Delta area, although fertile, is marshy and therefore not suitable for animals and growing crops (David, 1982). Gardiner (1961) wrote about the other fertile areas. These are the oases which are below sea level and are fed by springs and the Fayum which is the area around Lake Moeris.

The temperature is hot in Upper Egypt and temperate in the Delta and, as the rainfall is sparse, in antiquity the fertility of the land depended on the inundation caused by the rains from Central Africa raising the level of the Nile. The White Nile begins its four thousand-mile journey to the Mediterranean from the lakes of equatorial Africa and, when it reaches Khartoum, it is joined by another tributary, the rapidly flowing Blue Nile, which originates in the Ethiopian Highlands. The last tributary to join the Nile is the Atbara before it descends over the five cataracts (Stetter, 1993). The flooding reaches Aswan in Upper Egypt, by the end of June and arrives in Cairo, in Lower Egypt, by September. Prior to the construction of the High Dam at Aswan the water level gradually fell and was at its lowest level by the following April. The quantity of water which came down with the inundation was variable, but it is now controlled by the Dam. However, this is alleged to be silting up, and will soon be unable to control the water levels (Stetter, 1993). Previously, if conditions were good, the fertile silt which covered the banks of the Nile ensured an abundant harvest. However, if there was too much water, flooding destroyed the crops and, if there was too little, this resulted in famine.

Herodotus said that there were three theories, two of which he dismissed unconditionally, to explain the inundation. The third theory *'was the most plausible but at the same time furthest from the truth. According to this, the water of the Nile comes from melting snow, but as it flows from Libya through Ethiopia into Egypt, that is, from a very hot into a cooler climate, how could it originate in snow?'* In the introduction to *'The Histories'* (Selincourt, 1972), the Editor has stated that this is one of the occasions when Herodotus denied the evidence of available knowledge of that time as he did not believe the explanation.

2.2 Irrigation

The inundation had to be used to maximise the benefits of the fertile silt. To control the water levels, areas for cultivation were surrounded by earth walls and connected by canals. Because the water was held in the circumscribed areas it was thoroughly absorbed by the soil. If the level of the Nile was low it was possible to raise the water to the fields at a higher level for irrigation by means of the *shaduf*, which is used even now. This is a primitive device in which there are two poles with a hinged pole suspended between them. At one end is a rope with a bucket and this is balanced by a counter-weight (Emery, 1961)

2.3 Cultivation

Crops were sown in November, as the water level began to fall and harvested in April, before the next inundation. Barley *(Hordeum vulgare)* and emmer, a species of wheat *(Triticum dicoccum),* were the cereals most frequently produced. There was an abundant supply of vegetables such as lentils, beans, cucumbers, leeks, garlic, radishes and, in particular, onions. Fruits were also plentiful. There is evidence of grapes, dates, figs, and pomegranates (Gardiner, 1961). Much of the information about the diet has come from the contents of the tombs (Darby *et al.*, 1977).

- Mummified food offerings (including meat and fowl) and bread deposited for the afterlife.
- Food has been analysed which has been found in the viscera contained in canopic jars.
- Depiction of food on wall reliefs and paintings showing offering tables laden with a variety of foodstuffs.

  There are literary sources of information (Darby *et al.*, 1977).
- Papyri texts listing rations which were given to soldiers and workmen.

- Lists, in temples, of food offerings given to the gods.

There was also oil produced from the fruit of the ***b3ḳ*** tree which is now thought to be ben-oil from the moringa tree (Gardiner, 1961). Although there is no evidence of oil palms in predynastic Egypt, a fragment of the pericarp of an oil palm was found in a Mesolithic site in the Sudan which may mean that food substances were moved between regions. The olive tree did not appear until the 18th Dynasty. Flowers were also grown and a panel on an ivory casket in Tutankhamun's tomb illustrates the king and his wife in a garden surrounded by flowers. There is a wall painting in a 19th Dynasty tomb showing cornflower, poppy and mandrake (Manniche, 1993). The lotus was found in the swamps around the Delta and is seen in wall paintings of banquets. Garlands have been found around the necks of mummies. From the funerary temple of Queen Hatshepsut at Deir El-Bahari, there is an illustration of her expedition to the Land of Punt where incense trees were collected to be replanted in her gardens.

Flax was grown and was spun to fulfil the necessity of providing linen thread for clothes. In the earliest dynasties cloth was always made from vegetable fibres. Grass, reed and hemp were also woven. The linen was of several qualities. The finest was Royal linen. Next in quality was fine thin and then thin and smooth linen. Recycled household linen was given to the embalmers. Because of superstition, wool was not used for clothing which was to be worn in the temples or placed in tombs (Gardiner, 1961).

Papyrus also grew in abundance. Thin slices of its pith were laid at right angles to one another, beaten together and dried in the sun to make a medium suitable for writing. The word, paper, is derived from papyrus. Other parts of the plant were used to make rope and sandals. The root of the Papyrus plant was roasted and was used as a masticatory and this will be discussed below (Manniche, 1981).

There are many bas-reliefs showing corpulent figures and others illustrating scenes of famine (Darby *et al.,* 1977 & Nunn, 1996). Although there was an abundant variety in the food grown, there was much malnutrition caused by the variable water levels of the Nile. Other factors may have played an important part. Over consumption of cereals relative to other foods may have resulted in protein energy malnutrition (Brothwell, 1967). Some essential foods may have been in short supply and available to only a small percentage of the population. There may have been religious prohibitions on the use of some foods to the general population.

2.4 Livestock

There were herds of African cattle. Oxen were used for farm labour and also for sacrifice (Gardiner, 1961). Sheep, goats and pigs were kept and the goats and pigs were used to tread in the seed. Sheep were seldom sacrificed or killed for their meat as they were the representation of the god, Amun. There was a partial dietary prohibition of the pig. Geese and ducks were kept. River fish were netted and then dried or smoked. Bees were kept and the honey collected was used both as a sweetening agent and, as an ingredient, in many of the prescriptions in the medical papyri.

2.5 Diet

Most of the dynastic population of ancient Egypt lived on the banks of the Nile. The earlier predynastic populations from the Palaeolithic period lived on the edge of the desert as the Nile Valley was under water and the vegetation abounded with wild life. They were hunters, and their weapons, which would have been used to kill wild animals, have been found. There is more information remaining from the Mesolithic period. Near the village of Sebil in the Kom Ombo plain there is evidence of three different periods of occupation. They are known as the Sebilian culture (Saffiro, 1969). The Kom Ombo plain had originally been covered with water and, as the water dried up, the population had settled at successively lower levels to be near the water. The second and third period excavations have yielded millstones and stone pestles. There were also food refuse heaps in which were discovered remains of molluscs, fish bones and herbivore bones. The bones had been opened to extract the marrow. This population showed a transition between hunting and food production. In the north of Egypt on the east bank of the Nile opposite Abusir was the El Omari settlement. The site of El Omari may have been even earlier than that of the predynastic, El Badari (Midant-Reynes, 1992). There were the remains of granaries and dwellings lined with matting in which were abundant remnants of wheat and barley, a cake made of wheat and fragments of wheat and barley bread. There were sycamore figs and dates and stalks of wild sugarcane – saccharum spontaneum L. This is the earliest example of the plant recorded in Egypt. Among the animal remains found were pig, hippopotamus, crocodile, snail, ostrich, antelope, goat and fish bones (Hayes, 1964).

In the fertile region of Egypt known as the Fayum, the Neolithic population began the cultivation of cereals. Agriculture and stock farming were possible because of the Neolithic Sub-pluvial or Moist Interval when the extent of the grasslands was greatest. Coarse pottery cooking pots containing fish and animal bones have been found. There were silos with barley and wheat and axes and sickle blades of limestone and dolerite. There were fish spears made of fish bone and basket fish traps (Hayes

1964). It is thought that a rudimentary form of unleavened bread was made and baked on heated clay. Saffirio (1969) has suggested that their diet was the result of some agriculture and animal husbandry augmented by hunting, fishing and food gathering. Before 4000 B.C the predynastic people of Middle Egypt were known as Badarian. This was a culture named after the modern village of El-Badari. Their food was obtained mainly as the result of hunting and fishing but during archaeological excavations, by Brunton and Caton-Thompson in 1928, spoons and ladles were unearthed. According to Saffirio, this indicated that there was soup or porridge in the diet. The later predynastic periods, from 4000 to 3000 BC, were known as Naqada 1, 2 and 3. They were named after the site in Upper Egypt excavated by Sir Flinders Petrie (David, 1982). Pieces of bread have been found in some of the vessels of that period. At Abydos brick constructions were found which are thought to be for drying cereals to remove the husks from the grains before grinding.

Bones of animals which may have been hunted such as hyenas, gazelles and fish have been found in the Siwa and Kharga Oases together with drawings of giraffe, antelope, ibex and ostrich. There were also hand mills for grinding grain.

In the western delta saw-toothed sickle flints have been discovered (Hayes 1964).

At the beginning of the dynastic period there seemed to be a deterioration in the food supply because of an increase in the population (Saffirio, 1972). Darby *et al.* (1977) do not agree with this as they consider that the diet was supplemented by the addition of fruit, vegetables, dairy products and fish. The previous hunting and gathering lifestyle was exchanged for one dependent on agriculture and animal husbandry The diet consisted of bread, vegetables, fruit and fish. Cow's milk was part of the diet and there is an illustration in Darby *et al.*, (1977) of a child suckling a cow at the same time as the calf. Cheese may also have been manufactured as remnants of cheese have been found in a 1st Dynasty jar. It is thought that there was a reduction in the quantity of meat available as large animals were used as beasts of burden and for treading in seeds, resulting in a diet which was mainly vegetarian.

Another source of information about the diet is in the Old Testament (Hertz, 1976) which quotes the people of the Exodus who had left Egypt complaining that *'We remember the fish which we were wont to eat in Egypt for nought; the cucumber, and the melon, and the leeks, and the onions, and the garlic…'*.

2.5.1 Meat and fish

In the Predynastic Period, because the Neolithic peoples were hunters rather than agriculturists, meat was a major constituent of the diet. Deposits of animal bones, including those of the hippopotamus complete with arrowheads, have been found near settlements (Saffiro, 1969).

From dynastic times meat would only be available to the wealthy. The priests would have been responsible for the sacrifice of cattle and would have had access to beef as food (*vide infra)*. Royalty, priests and the upper echelons of society would have consumed most beef (Darby *et al.*, 1977). There are many tomb paintings showing butchery and models of butchers' shops from the Old, Middle and New Kingdom tombs are to be seen in museum collections. There is evidence that, in addition to meat, the brains were also consumed as, in a bovine cemetery, skulls were found split. Preserved beef has been found in tombs. The meat was cooked mainly by boiling as it was thought that it allowed the evil to escape. They may have believed that roasting trapped the evil in a hard crust. Ox grease was used in prescriptions for coughs or, externally, for the treatment of burns (Darby *et al.*, 1977).

Goats and pigs were used to tread in the seed but seemed, at least partially, to have been forbidden as food although there is some evidence that pork *was* part of the diet in some areas. This may have been influenced by the worship of Seth who was sometimes represented as a pig, or in the cult of Horus in which the pig is an abomination and not used in sacrifice (Hart, 1986). In predynastic settlements at El Maadi bones of swine have been found indicating that they were eaten (Darby *et al.*, 1977). During the examination of Nakht, a weaver from the 20th Dynasty, now in the collection at the Royal Ontario Museum, a cyst, thought to have been caused by the parasite, trichinella, was found. This cyst would have formed as the result of eating inadequately cooked pork (Millet *et al.*, 1985). Nakht showed signs of malnutrition or chronic illness, as Harris lines were present in his long bones. However, when Herodotus visited Egypt in the 5th Century B.C., he said that '*the pig is regarded among them as an unclean animal, so much so that if a man in passing accidentally touch a pig, he instantly hurries to the river and plunges in with all his clothes on.'* And '*swineherds, though of pure Egyptian blood, are the only people in the country who never enter a temple nor is there any intermarriage between them and the rest of the community, swineherds marrying their daughters and taking their wives only from amongst themselves.'*

There is evidence of the presence of sheep in Egypt from predynastic times but there was a prohibition on the priests from using the wool and eating the meat because of the worship of the ram god, Khnum. Herodotus said of the common man that '*the clothes they wear consist of a linen tunic with a fringe hanging round the legs and a white woollen garment on top of it. It is, however, contrary to religious usage to be buried in a woollen garment, or to wear wool in a temple.'*

During the examination of Natsef-Amun, a priest from the Temple of Karnak, in the 21st Dynasty atheromatous plaques were found in his femoral arteries. This meant that he must have had a significant cholesterol intake (Tapp & Wildsmith, 1992). The explanation for this finding is that a priest would have had access to the daily food offering for the gods of the temple for which sacred cattle would have been sacrificed. The priests removed the food at the end of the daily ritual from the sanctuary to the temple precincts where it was divided up as part of their wages (David, 1992).

Eggs from the geese and ducks were plentiful but it is not known whether they were a part of the diet or whether they were used only for the continuation of a supply of the animals themselves by hatching. Eggs were also used in medicines (Darby *et al.,* 1977).

Fish from the river were readily available. They were gutted and dried in the sun (Herodotus, 1972)

2.5.2 Wine, beer and bread

Wine was made from grapes and pomegranates. Beer and bread are recorded as dietary staples and have been discovered as offerings in tombs and are depicted as loaves and jars of beer on wall decorations particularly in Old Kingdom tombs (Samuel, 1993). Analyses have been made of the dregs of beer and bread found in the tombs. Forty different kinds of bread and seventeen beers have so far been identified. For example, there is a red loaf in the collection at the Museum of Fine Arts in Boston in which one of the main ingredients has been identified as fig. Bread moulds have been discovered in the New Kingdom excavations at Amarna, in temple bakeries. Their different designs have been compared with moulds from Old Kingdom and Middle Kingdom sites. Analyses of the grains found at various sites have been carried out and barley and emmer, a coarse wheat, adulterated with weed seeds, have been recognised. One of the techniques for beer making has been found in tomb scenes. In this, it is seen that after the bread was baked, it was soaked, allowed to ferment, and then strained. Bread was of such importance in the Egyptians diet that they were known as 'eaters of bread' or 'artophagoi'.

Samples of bread from several museum collections have been analysed (Leek, 1972). These bread samples, from 3000 B.C. until 1085 B.C., were from different sites. They were compared with *fellahin's bread* eaten by excavators during an excavation, 'village bread', and English stone ground bread and white bread. Almost all of the ancient samples appeared to have whole grains on the surface and inorganic contaminants. When the fragments were softened the whole grains were found to be husks. However this is because the endosperm had degenerated with the passage of time. The samples were then examined by x-ray with 4x magnification to show the inorganic contaminants more clearly. After treating the specimens to isolate the particles, they were photographed and magnified x 105 and compared with samples of sand from sites in Egypt. The particles were identified in the Department of Mineralogy at the Natural History Museum. Most were rounded grains of quartz, which were recognised as sand but grains of feldspar, mica and hornblende were also present. Only the samples of English bread had no inorganic contaminants. Leek (1984) cited Pliny the Elder (23-79 AD) as having written that the Carthaginians added sand and chalk to the grain to aid the crushing of the grain. Prag, of the Manchester University Museum, in a personal communication to Leek (1984), told of his experiments using an ancient quern to repeat this technique. He was thus able to produce fine flour. Previously, he had pounded the corn for fifteen minutes with no appreciable change. The Egyptians may have applied the same technique but it is also possible that the accidentally incorporated particles may have had the desired effect in breaking down the corn. These particles came from soil itself and from the flint-tooth sickle harvesting tools seen in 18th Dynasty tomb paintings. Wind-blown sand may also have been incorporated during winnowing. During grinding the surface of the querns and saddle-stones were abraded over time and would have made their contribution. There are also illustrations of stone ovens with bread baked on the outside where the surface of the bread might have been further adulterated by sand and granules of stone.

The effect of the bread contaminants on the dentition will be discussed (*vide infra* chapter 3).

2.5.3 Fruit and vegetables

Fruit and vegetables were the main components of the diet and were enjoyed and eaten in abundance. Herodotus reported about the diet of the pyramid builders that *'The finishing-off of the pyramid was begun at the top and continued downwards, ending with the lowest parts nearest the ground. An inscription is cut upon it in Egyptian characters recording the amount spent on radishes, onions, and leeks for the labourers, and I remember distinctly that the interpreter who read me the inscription said the sum was 1600 talents of silver. If this is true, how much must have been spent in addition on bread and clothing for the labourers during all those years that the building was going on------'*

2.5.4 Honey and other sweetening agents

Food was sweetened with honey, dates and figs.

Honey (***byty***) was first gathered from wild bees in the Delta. This was a luxury and, as it was very costly, only the wealthy could afford it. In the earliest period it was the king's prerogative and was used for religious purposes only. By the Middle Kingdom, it was found in private tombs as an offering to the gods. Later it became more attainable for the general population.

Honey was incorporated in cakes and pastries (Darby *et al.,* 1977). It was also used in medicines and its use in dental prescriptions has been described in Chapter 1 in the *Edwin Smith Papyrus* and the *Ebers Papyrus*. Honey contains two sugars, fructose and glucose.

There have been finds of dynastic dates and date-cakes (Darby *et al.,* 1977). Dates are composed of 75% glucose and fructose and were also used to make wine and sweeten beer. Palm wine was used to wash the body during mummification (Brewer *et al.,* 1994 [1].

As mentioned in 2.5.2 a specimen of bread in the Museum of Fine Arts in Boston is reddish in colour. Samuel (1993) identified fig seeds which gave it its colour and sweetened the bread.

Fruit juices were also in evidence as sweetening agents. These were used mainly by the population who were unable to afford honey. Date juice was added to sweeten beer (Darby *et al.,* 1977). In the Cairo Museum is a jar with a label '*unfermented grape juice of good quality from the temple of Aten*' from Tutankhamen's tomb.

2.5.5 Masticatories

Although these are not strictly a part of the diet, their use had an effect on the dentition and they are therefore included in this section. Masticatories are substances which are chewed but not ingested. The betel nut from Asia, which was used historically and is still chewed today, is an example (Dixon, 1972). A modern, almost universal, masticatory is chewing gum. Masticatories are expectorated and their remains, called quids, have been found at archaeological sites. There are three distinct types of masticatory which seem to have been in use in Egypt and there is information about all of these in contemporaneous writings.

Herodotus wrote that the Egyptians used the root of the papyrus, which they roasted in a red-hot oven. It was then chewed, the juice extracted and the quid spat out. When examined, the quids often contained phytoliths, inorganic impurities, from the soil which would have had an abrasive effect on the teeth.

Natron pellets were chewed to refresh the mouth for both religious and social reasons. Natron is a naturally occurring compound found in low-lying areas of Egypt, which become flooded during the inundation bringing the deposits to the surface. It is a mixture of sodium carbonate, sodium bicarbonate and a small percentage of sodium chloride and sodium sulphate. As the pH is high, it neutralises the low pH of the saliva which results after eating carbohydrate and which causes demineralisation of dental enamel. In the Pyramid Texts it is said that '*the mouth is like that of a sucking calf on the day it was born'*.

In the *Ebers Papyrus* there are several preparations, medicaments, which were used as masticatories. These are described as ***ḫpꜥw*** which comes from the verb 'to chew', but the papyrus stated that after being chewed the substance was expectorated.

2.6 Dietary causes of disease

The most ancient Neolithic community in Egypt is the Fayum group. C_{14} has been used to give a date of 4440 BC +/- 180. The 25aemorrha population who had become agriculturists or pastoralists greatly increased in size and was dependent on the cultivation of the fertile land around the Nile. Because the people were now living in close proximity they were more susceptible to epidemics of disease which increased early mortality. The production of the food was conditional on the inundation. From the 3rd Dynasty there is a report of famine during the reign of King Djoser, '*the Nile has not risen in my time for the space of eight years. Corn is scarce, there is lack of herbage, and nothing is left to eat.'* There were accounts also of cannibalism during famine from a later period and it may have occurred in predynastic times (Darby *et al.,* 1977). There are many illustrations of emaciated figures, such as for example, Unas' ramp from the 5th Dynasty (David & David, 1992). This has been interpreted as a 'famine relief', showing a group thought to be dying of starvation and who are not Egyptian. It possibly indicates that they had received relief from Unas for their hunger.

2.7 Malnutrition (Protein-Energy Malnutrition)

It is possible that many people suffered from Protein-Energy Malnutrition (P.E.M) or Kwashiorkor which potentially led to a destruction of the periodontal tissues. Kwashiorkor, which is P.E.M. in childhood, was first described in Ghana in 1933 and is known to occur in many of the underdeveloped countries. It is severe protein deficiency and is associated with a low calorific intake (Dunlop & Alstead, 1966, Maegraith, 1989). It comes from the Ghanaian word which describes the *disease suffered by a child displaced from the breast* (Golden, 1996). It is associated with scaly skin and discoloured, often orange hair, called '*dyspigmentation*' (Cecil & Loeb, 1960[2].

Porter (1997) has written that, '*Agriculture allowed more mouths to be fed, it meant undue reliance on starchy cereal monocultures like maize, high in calories but low in proteins, vitamins and minerals; reduced nutritional levels allowed deficiency diseases like pellagra, marasmus, kwashiorkor and scurvy to make their entry onto the human stage. Stunted people are more vulnerable to infections.*' The consequence of this is seen in the Neolithic skeletons which are shorter than their Palaeolithic antecedents.

In malnutrition, there is deficiency of the trace elements zinc, copper and also of vitamin A. Chandra (1993) wrote that zinc deficiency results in life threatening complications of infections. It may also be responsible for skin lesions. Lack of copper is associated with increased incidence of infections. The hair is sparse (Southami & Moxham, 1990). There is also gingival inflammation and periodontal bone loss. In animals with protein deficiency there are reduced deposits of cementum, degeneration of the periodontal ligament and delayed wound healing. However, true periodontal pocket formation does not occur in the absence of plaque (Wilson & Kornman, 1996). In experimental animals there was degeneration of gingival tissue, the periodontal membrane and osteoporosis of the alveolar bone (Shafer *et al.*, 1958[1]. Food may be contaminated with plant fungi which affects the liver and pancreas and may also be the cause of the skin and hair changes (Maegraith, 1989). If there are bouts of malnutrition there are bands of discolouration in the hair known as the *flag sign.* This might be equivalent to the formation of the Harris lines in the long bones and also to dental hypoplasia.

Marasmus, also found in children, is the result of starvation. The diet may be nutritionally balanced despite there being a severe shortage of calories. Shafer *et al.*, (1958[2] wrote *'no specific oral lesions have been described'.* This is confirmed by Bradford (1958) who thought that there might be a racial predisposition to periodontal disease and that people who have this problem live in the tropical or subtropical areas and depend on cereals for food. This is the very group of people who are found to have severe protein deficiency. Nomadic people from the same area do not seem to experience P.E.M and periodontal disease because there is animal protein in the diet.

The periodontal membrane is a fibrous joint with collagenous fibres running from the cancellous alveolar bone to the cementum which covers the root of the tooth. Unlike other joints of the body it is directly exposed to the oral environment at the gingival margin. The vascular cancellous bone is affected by environmental changes more than the cementum. The local effect of the diet initiates the inflammation of the gingival tissues and the subsequent loss of the attachment of the epithelium around the neck of the tooth which allows the ingress of the oral bacteria (Bradford, 1958). This probably occurred rapidly in this population because the bread from the emmer wheat was very sticky and would have adhered around the gingival margins. There is no knowledge of any oral hygiene regimes in the ancient Egyptians except for the occasional use of natron pellets masticatories (2.5.5). The progress of the periodontitis begins with the inflammation from the gingivae affecting the periodontal ligaments and the alveolar bone. Collagen degradation, bone resorption, and pocket formation follow this (Soames & Southam 1998).

The vomiting associated with P.E.M. might be responsible for the erosion found in the dentition of some of the skulls. Because there is an associated deficiency of trace metals in protein deficiency malnutrition it could also be the cause of the red headed mummies. 'Ginger' in the British Museum was found in a predynastic sand grave and has red hair. One might speculate whether the hair colouration was natural, caused by 26aemorrhage change in the hot dry environment or reflected changes from malnutrition. P.E.M. is associated with gynaecomastia with gross hepatomegaly because of fatty degeneration (Golden, 1996). Hepatomegaly is also seen in sufferers from schistosomiasis (Mowat, 1987). This latter disease might explain the artistic representation of the New Kingdom pharaoh, Akhenaten. It is unlikely that he would have suffered from nutritional deprivation.

2.8 Scurvy

Another cause of periodontal disease is scurvy. Its aetiology is the lack of vitamin C that is fundamental to the biosynthesis of collagen. Its lack will cause haemorrhage under the periosteum, skin and bleeding into the joints. When the bleeding organises, new bone is formed on the alveolar bone, orbit and the attachment of the temporalis muscle (Roberts & Manchester, 1995). There is also hyperkeratosis of the hair follicles, loosening of the teeth and subcutaneous haemorrhages (Maegraith, 1989). The connective tissue of the periodontal membrane is affected and there is often a secondary fuso-spirochetal infection (Lynch, 1977). There are oedematous changes in the gingival tissue with haemorrhage into soft tissues. The teeth become mobile with the destruction of the periodontal ligament and osseous abnormalities (Wilson & Kornman, 1996). It is possible that there was a lack of vitamin C in the diet in the predynastic communities when they became settled and dependent on their staple crops. Experimental work found that volunteers remained in good health with an intake of only 10mg. per day but British nutritionists suggest 30mgs and American nutritionists advise 75mgs as optimum (Cecil & Loeb, 1960[2]. Fresh fruits, vegetables and raw meats contain ascorbic acid. Vitamin C is found in cabbage, cauliflower, tomatoes and citrus fruit. The only citrus fruit grown in Ancient Egypt was the citron. All other citrus fruits were introduced at a later date (Darby *et al.*, 1977).

Rich sources of vitamin C are found in blackberries, melons, and pomegranates which were grown in ancient Egypt, whereas there is little vitamin C found in dates and figs, which were used to sweeten foods. Artichokes and asparagus, a rich source, were a late introduction to Egypt and cabbage, which was found in a tomb by Petrie in Hawara, was thought to counteract the effect of wine (Darby *et al.*, 1977). *'It is an advantage to include in the menu cabbage boiled with meat; olives boiled in water, and the like. For this conduces to drinking more wine....'* Celery is found on mummies in tombs. The *Ebers Papyrus* refers to its use in medicaments. Cress and parsley were grown and are a rich source of vitamin C. Other vegetables such as onions, cucumbers, garlic, leeks, lettuces, pumpkins, radishes and peas also contain small quantities. Ebbell (1937), in his interpretation of *The Papyrus Ebers,* translated 'blood eating' as scurvy but there is no evidence that this is an accurate paraphrase of the Hieratic.

2.9 Hypocalcaemia

This is a condition that could have occurred during famine caused by drought or flooding. If the individual is undernourished and has suffered from trauma, the calcium loss will result in demineralisation of the skeleton and pathological fractures. (Cecil & Loeb, 1960[2].

The fertility of the Nile Valley and the development of irrigation systems gave rise to the cultivation of an abundant variety of crops and breeding of animals. The food supply was excellent and varied but the periods of famine had an adverse effect on health and the dentition. The following chapter discusses the history of dentistry in Ancient Egypt.

Chapter 3

The Title of Dentist in Ancient Egyptian History and Conflicting Evidence of the Practice of Restorative Dentistry

3.1 Dental practitioners in Ancient Egypt

Herodotus said '*the practice of medicine they split up into separate parts, each doctor being responsible for the treatment of only one disease. There are, in consequence, innumerable doctors, some specializing in diseases of the eyes, others of the head, others of the teeth, others of the stomach and so on; while others, again, deal with the sort of troubles which cannot be exactly localised.*'
There is evidence of almost one hundred medical practitioners and six of these had the title of dentist (Weeks, 1980). Several titles were used: ***ibḥy,*** which is translated as 'one who is concerned with the teeth', is derived from the word ***ibḥ*** – a tooth and ***iry ibḥ,*** who is 'one who treats teeth'. Though these titles differ they define the role of a dentist. In their context in ancient Egyptian writings they refer to someone who was subordinate to a physician. ***Wr ibḥy*** and ***wr iry ibḥ*** are both translated as 'Chief of Dentists' but their difference is unclear. The 'Chief Dentist of the Palace', ***wr ibḥy pr-ꜥ3,*** is the most senior title that has been so far discovered.

Of the six named dentists, five were from the Old Kingdom from the 3rd to the 5th Dynasty, and only one is known from a later era, the 26th Dynasty in the Late Period, an absence of the title for almost 1700 years. It is unlikely that the title was unused for that length of time and so further discoveries might still be made.

The identified dentists are:

- **Hesyre** is the earliest named dentist. Several wooden panels remain from his 3rd Dynasty tomb. On the top right register of one are the Hieroglyphs which, in transliteration, read ***wr ibḥ swnw***. The ***wr*** means 'great one', ***ibḥ*** is the tusk sign and ***swnw*** is translated as 'doctor'. This is interpreted as 'Chief of Dentists' and 'Chief of Physicians'. Also, among his titles are 'Overseer of King's Scribes', Overseer of Draughtsmen', 'Elder of Builders', 'Chief of Taxes and 'Father of Min', a priestly role (Rawlings, 1999). There are several other titles on the stele. It is possible that these are all ceremonial and he had no active role in tooth care. However, it may acknowledge the necessity for dental support.
- **Neferiretes** had the title of 'dentist'.
- **Khuwy** was a very senior practitioner, who was not only 'Chief of the Dentists' but also 'Chief Physician of Lower and Upper Egypt'. In addition he was 'Doyen of the Physicians of the Palace' and 'Guardian of the Anus'. He treated the gastrointestinal tract at both extremities!
- **Redi en Ptah** was a Physician and an Overseer of Physicians and an Overseer of Dentists (Ghalionghui 1987).
- **Nyankhsekhmet** was another very distinguished practitioner in the 5th Dynasty who was 'Chief of the Dentists of the Palace', 'Chief of the Physicians' and 'Chief of the Physicians of the Palace'.
- **Menkaureankh,** also in the 5th Dynasty, had the more humble title of 'dentist'.
- It was not until **Psammetik-seneb**, in the 26th Dynasty, that the title has been rediscovered. He was 'Chief of the Dentists of the Palace', 'Chief of the Physicians' and 'Doyen of the Physicians'.

3.2 A dental profession?

Since the advent of the study of forensic Egyptology, it has been a matter of dispute whether any interventional dentistry was carried out in ancient Egypt or, if the practice of dentistry was limited to the use of the prescriptions in the medical papyri. What is not in doubt, from the skeletal evidence, is the extensive and severe dental disease suffered by the ancient Egyptians. As the title of 'dentist' was used in the Old Kingdom, and the medical papyri were in existence from the Middle Kingdom onwards, it is astonishing that there is no significant evidence of the advanced technical skill that the Egyptians had developed in other fields being utilised for the relief of pain and the restoration of function and appearance in the ravaged dentition.

3.3 Dental disease

The necessity for dental intervention was obvious from the Old Kingdom as can be seen from the use of the titles. It also is apparent from the evidence of severe dental disease seen in the mummies and skeletal material from these earliest times. This will be illustrated in later chapters. The medical papyri so far discovered are from

1950 B.C. until 1490 B.C., the late Middle Kingdom until the New Kingdom, and as was discussed in Chapter 1, were based on earlier treatises.

The dental conditions documented in previous research were caries, attrition, hypoplasia, periodontal disease, periapical lesions, osteomyelitis and osteoarthritis of the temporomandibular joint. The causes of these problems can be explained by the geography of Egypt, the diet and the social conditions present during the four thousand years and these have been investigated for this investigation.

3.4 Dental disease experienced by the Ancient Egyptian

From the evidence of previous examinations of skulls, caries was not the problem that it is today. Some of the periapical lesions which the ancient Egyptian developed were caused by caries but were mainly the result of the relentless attrition / abrasion which wore away even the secondary dentine and allowed the ingress of bacteria into the pulp chamber. The site of the caries, which was prevalent in the predynastic period, was predominantly in the cervical region of the tooth. The incidence of caries decreased in the dynastic period from 33.33% to 28.1% of individuals examined and from 6.14% to 4.65% of the teeth examined (Grilletto, 1973). Because many teeth had been lost both ante- and postmortem these figures might be inaccurate. The postmortem tooth loss in the examined skulls has been found to be mainly in the anterior segment of the dentition because these teeth have single, conical roots and, once the periodontal ligament disintegrates, they are easily dislodged. These teeth are also the least susceptible to caries and so the percentage of teeth affected may be much less than the previously recorded figures.

3.5 Aetiology of caries

The aetiology of caries was thought to be a 'worm'. There is the '*Legend of the tooth worm'*, which is a magical text of the Babylonians, in which toothache is caused by a demon, represented by a worm, which must be destroyed to cure the pain. Maggots found in figs were analogous to the tooth worm (Weinberger, 1948).

An even earlier example of this is found in Papyrus Anastasi IV (1200 B.C.) which quotes an Egyptian official thus; '*I spend the whole day watching the birds--- There is the gnat at sunset and the midge at noon; the sandfly stings and sucks at every vein-----A* ***mns****-scribe is here with me, every muscle of whose face twitches, the* ***wštt****-disease has developed in his eye and the* ***fnt****-worm into his tooth. I cannot leave him to his fate.'* The ***fnt***-worm is translated as a real or metaphorical intestinal worm' (Nunn, 1996).

3.6 Evidence of dental treatment

3.6.1 A bridge from the 4th Dynasty

The first evidence was found by Junker in 1914, in a mastaba tomb at Giza which dated from the late 4th or early 5th Dynasty. He decided that a 'bridge' had been made by the embalmers during mummification (Leek, 1984). A length of gold wire (Weinberger, 1947 and Leek, 1967a) united an upper second and third molar. The third molar is reported to show more attrition than the second and its roots have been almost completely resorbed. Professor Euler, the Director of the Breslau Dental Institute who examined the teeth for Junker, believed that the teeth had belonged to the same individual because of the similarity in their colour and anatomical form. He considered that the third molar had become mobile because of its pathological condition and was held in place by attaching it to the second molar in a method which, he said, was still carried out in 1928 when the teeth were examined! He also stated that 'tartar' on the wire proved that the procedure had been carried out in the mouth of a living person. The photograph (Figure 1) shows the teeth wide apart but Weinberger (1947) wrote that, in Junker's report, both teeth were said to be close together as is usually found in the mouth. Ghalioungui (1963) also sites the bridge as evidence of dental treatment. Leek (1967a) was sceptical about Euler's conclusions. He had, however, only examined the photograph and had not seen the appliance. Leek reported that the only method by which the concretions on the wire, which were called 'tartar', could be verified was if the sample was examined microscopically. He explained that there was often an accumulation of material, resembling calculus, on bones, which had been buried over a long period. However, the prosthesis must have been interred separately as there was no evidence of a skull present in the mastaba and so the 'bridge' was unlikely to be in its owner's mouth at death. The wire is tied between the teeth but, the most convenient position to tie the wire in the oral cavity would have been mesial to the second molar. The attrition is much greater on the third molar than on the second molar. This is improbable, as the third molar would have erupted approximately six years after the second molar and would not have been subjected to as much wear because of its age and anatomical position. Filer (1995) stated that Schultz of Gottingen University has suggested that the teeth were not from the same dentition and that Schultz was hoping to examine them microscopically.

3.6.2 Another bridge from the 4th Dynasty

Harris, Iskander and Farid (1975) described a bridge which was found in 1952 by Farid in a mastaba which dated back to the 4th Dynasty. The bones of the skeleton were almost completely crushed by a fall of stone and there is no information whether there were any other teeth present. The specimen consisted of the upper right canine, with an intact root, which has a double wire wound around its neck and tied at the distal aspect. This was considered, by the authors, to be the abutment. Adjacent to it is a lateral incisor with a groove cut into the labial surface to accommodate the wire which was then tied around the tooth. The root appears to have been modified intentionally as it is both shortened and narrowed. The central incisor also has a labial groove with a hole drilled mesio-distally to accommodate the wire. The root of this tooth also appears to have been altered. It is suggested that this was attached, *in vivo*, to the left, central incisor which was the second abutment to the bridge. Again, Harris, Iskander and Farid found calculus on the canine and lateral tooth which, they considered, confirmed that the bridge had been worn for a long period of time. Weeks (1980) alleged that the cemetery had been reused in Ptolemaic times and so the date of the bridge was speculative. This might advance the date of the construction of the bridge by two thousand years.

Another specimen, which has been found in Phoenicia (modern Lebanon), has gold wire tied around the gingival margins of the lower incisors and canines to support these teeth. There is also a Roman appliance, made of rolled gold, extending from the lower left lateral incisor to the lower right canine. It is 3mm in height and is contoured to the gingival margin. Weinberger (1947) suggested that the later examples were probably of Egyptian construction.

3.6.3 Root canal treatment?

Zias and Numeroff (1987) examined skulls from a mass grave in the Negev Desert which dated from Graeco-Roman period, approximately 200 BC. One skull had a discoloured, stained green, right maxillary lateral incisor. A radiographic examination showed that the root canal had been widened and a 2.5mm bronze wire was inserted into the canal. It was suggested that this had been placed in the tooth to prevent the penetration of 'tooth worms' which were considered to be the cause of dental pain. Another theory is that it had been placed to help to drain the large apical cyst. Zias and Numeroff (1987) also suggested that, because of bronze having a similar colour to gold, it had been substituted as a fraud, despite the knowledge of the corrosive nature of the bronze.

3.6.4 Extraction of teeth

There has also been much controversy about whether there was surgical intervention to extract teeth or to drain apical abscesses. Because of the loss of supporting bone caused by periodontal disease, and periapical destruction resulting from pulp death from attrition, there are many examples of teeth which would have been so mobile that they could have been removed by digital manipulation. These have been left in situ as in Figure 2a.

Dawson (1929) quoted from the Coptic medical papyrus of Meshaikh from the 9th Century AD which is from the time of Roman rule in Egypt. *'A tooth to be extracted with instruments. Hellebore of good quality and gall; apply to the region of the cheek where the molar is that you wish to extract, and you will be astonished.'*

Leek (1966) said *'In no instance was any interruption of the pathological sequence noted, and it is concluded that no human intervention other than the use of magical incantations took place for the relief of dental disease. It cannot be completely discounted, however, that a method of removing teeth was known.'*

Leek was very dismissive in the above statement about the possible historical dental intervention as there is the evidence of the dental prescriptions in the medical papyri which were described in Chapter 1. These were not exclusively magical incantations. Some may have been effective as antiseptics and others may have supported mobile teeth when the medicaments set to form a splint.

Figure 2b is an illustration shown by Miller (1997) of a possible dental extraction. There is no indication of periodontal disease in the 2nd and 3rd molar and the site of the missing 1st molar shows bone regeneration and no evidence that this tooth had exfoliated because of disease. This was the only specimen found. No instance was found in this survey.

3.6.5 Drainage of abscesses

Breasted (1930), in his foreword to his facsimile of the *Edwin Smith Papyrus,* said that the treatise mentioned the use of the 'fire-drill' for cauterisation. He suggested that this might have been used in a 4th Dynasty mandible to drain an abscess under a molar tooth. He identified what he thought was a drill hole in the mental foramen which, he surmised, had been made with a bronze drill. Although this technique had not been mentioned in the papyrus, Breasted considered that the technique must have been taken for granted by the author of the papyrus! Weinberger (1947), after he had consulted Thoma and Hooton, who had originally examined the mandible and affirmed their original opinion, wrote, *'Therefore there is not the slightest doubt that some individual in Egypt actually had had an operation performed for the purpose stated.'*

Ghalioungui (1963) agreed with Breasted that drilling holes was performed to relieve abscesses. Leek (1967b) attested that there was no necessity to drill a hole in the bone to drain an abscess because the expansion of the pus, under pressure, forms a sinus and perforates the compact bone of the alveolus. There are many examples of these holes (Figure 9), but no evidence has been found of any suitable instrument such as a right-angled drill which would be necessary to make the hole surgically.

However, when Bennike and Fredebo (1986) examined bones from Langeland from the Middle Neolithic period (3200-1800 BC), they found an intact skull with a hole drilled between the buccal roots of the upper second molar. Another hole had been drilled between the first and second molar where there was evidence of caries. Archigenes and Galen (100-200 AD) had said that if a tooth turned black it meant that the pulp was necrotic and that gases, which were the result of putrefaction, caused pain. They suggested that the tooth should be drilled to reduce the internal pressure. Bennike and Fredebo experimented with a Stone Age flint drill and made a similar hole in $5^{1}{}_{2}$ minutes. Both holes were examined with a scanning electron microscope and were similar, but the Stone Age tooth had evidence of 'calculus' in the drilled hole, and the authors concluded that it had been made during the life of the victim.

3.6.6 Filling of teeth

Ghalioungui (1963) suggested that teeth were filled and he quoted translations of the *Ebers Papyrus* in which he said that resin, Nubian earth and chrysocoll or Nubian earth and honey were used. In his unpublished translation of 1987 he deciphered these as substances to be used to 'consolidate a tooth'. No verification of this has been reported either in dental evidence or alternative interpretations of the papyrus.

3.7 ?Lack of evidence of dental treatment

Dawson (1929) said '*in spite of oft-repeated statements to the contrary, there is no evidence whatever that tooth-stopping, or any other form of dental surgery, was practised by the Egyptians in Pharaonic times.*'

It is difficult to imagine what kind of practical dentistry could have been practised in view of the lack of evidence from the many thousands of skulls examined. The title of 'dentist' may have been recognition of the severe problems of the inhabitants. However as the ***ibḫ***, the elephant tusk sign, which is claimed to denote 'dentist', disappeared at the end of the 6th Dynasty possibly the translation of the title as dentist may be inaccurate. Treatment may have been confined to the medicaments mentioned in the *Ebers Papyrus*.

Leek (1984) was of the opinion that '*although evidence has been produced by archaeologists and philologists and accepted by some medical historians as proof of the existence of an organised dental profession during the Old Kingdom, this evidence is difficult to accept after a detailed examination. Certainly it is not in accordance with the findings by examination of the human remains.*'

3.8 Evidence of surgical intervention

The only positive form of treatment which I came across was in that quoted in the Edwin Smith Surgical Papyrus which I have mentioned in this study in Chapter 2. This was for '*The reduction of the dislocation of the mandible*' For it to be mentioned, it must have been a common occurrence and was treated by '*the only procedure possible and so it is still used today.* (Miller 1997)' The technique is also described by Apollonius of Kitium in the 1st Century B.C. in his '*Illustrated Commentary on Hippocrates*'. What was the aetiology of frequent dislocation of the mandible? This will be discussed later. On the verso of the papyrus are spells and recipes. These are similar to those in the other medical papyri which do not include an examination of the patient but contain only the diagnosis and the treatment.

The technique is also described by Apollonius of Kitium in the 1st Century B.C. in his '*Illustrated Commentary on Hippocrates*'. The drugs used will be discussed later.

On the verso of the papyrus are spells and recipes. These are similar to those in the other medical papyri which do not include an examination of the patient but contain only the diagnosis and the treatment.

On balance, the conclusion reached from the evidence of earlier research is that dental treatment only became a reality during the Graeco-Roman Period.
The next chapter explores details of previous dental examinations of ancient skeletal and mummified material.

Chapter 4

Previous Dental Examinations of Ancient Skeletal and Mummified Material

Using procedures that depended on the condition of the human remains available and their site of storage, much information has been collected about the dental health of ancient Egyptians. The material was either skeletal or mummified. The mummified remains may have been unwrapped on some previous occasion or were still wrapped. In recent investigations mummified remains are left intact and often in their coffin. Such examinations must now be essentially non-destructive. Descriptions of the previous examinations performed on these various remains will be reviewed.

4.1 The examination of the dentition in skeletal material

The least complicated technique for acquiring data and, arguably, the most accurate, is the direct examination of the dry, clean skulls which are found in museum collections.

The advantages of this method of examination are:

- The environment in the museum is stable and there is usually sufficient space, light and time for a careful evaluation of the specimen.
- The presence or absence of teeth is easily observed. Bone regeneration in the sockets can establish whether teeth have been lost ante- or postmortem. . Postmortem loss results in clean sockets with sharp, alveolar bone margins.
- It is possible to observe and grade the degree of 'attrition' or wear (*vide infra* Molnar, 1972).
- It may be possible to identify the differences between the destruction of the alveolar bone caused by periodontal and periapical lesions.
- The condition of the disarticulated temporomandibular condyle can be classified by the change in form and state of its surface as an indication of osteoarthritis (Wedel *et al.*, 1978 - *vide infra*).
- Caries may be noted and are distinguished from post-mortem fracture of enamel.
- Age can be assessed in the young by the stage of development of the dentition. Consideration must be given to the variation in eruption time with girls' dentition tending to erupt earlier than that of boys. There is divergence in different ethnic groups (Hillson, 1996). There might be delay in the eruption times caused by chronic poor health. In the mature dentition the degree of attrition (Miles, 1962) is useful for estimating age. It may be necessary to consider the socio-economic group of the individual since this influences the diet and the effect of diet on the wear of the teeth.

Anomalies such as absent, retained deciduous, malformed, partially erupted and impacted teeth are evident.

The disadvantages are:

Museum material may be friable and may have been handled frequently. This might have resulted in postmortem loss of teeth and calculus or fracture of the enamel.

There may be what appears to be an alveolar fistula. This situation is not always an indication of a periapical lesion but the result of damage to the thin layer of cortical bone which covers the roots of teeth.

During the excavations mismatching of maxillae and mandibles by archaeologists may have occurred. Mismatching may also have occurred in museum collections.

Radiographic equipment is seldom available to confirm the assessed findings and therefore some pathology will be missed.

4.2 Gender

Gender may be evaluated to a greater degree by the landmarks of the skull and to a lesser extent by the dentition. These assessments will now be described in detail.

4.2.1 Gender from skull landmarks

The most accurate method of assessment of gender is by examination of the pelvis but skulls, also, show sexual dimorphism. Because excavators often discarded the rest of the skeleton in favour of the skull, there are seldom post-cranial remains available in the museum collections and so the skull alone is usually appraised. There are several landmarks that can be utilised to help to estimate the gender. Mann (1989) suggested that considerable discrepancies were found when he examined skulls which had been recorded in the excavation reports. There is also a significant overlap in the range of features found

Table 3: Differences between male and female skulls

	MALE	FEMALE
General size	Large, rugged and elongated	Smooth, small and rounded.
Supraorbital ridge.	Large	Small
Mastoid process	Large and straight	Small and curved inwards
Occipital area	Muscle lines and protuberance marked. Large condyles.	Muscle lines less marked, small condyles
Orbits	Squared with rounded margins	Rounded with sharp margins
Zygoma	Heavier and more laterally arched. Root extends beyond ext. auditory meatus.	Lighter and more compressed. Root stops at auditory meatus.
Palate	Larger, broader and U-shaped.	Smaller and tends to parabola
Teeth	Large	Small
Mandible	More robust, broader and longer ramus.	Smaller and lighter.

between the two sexes with a bias in favour of males (Brothwell, 1994).

Keen (1950) found, in his study of differences between male and female skulls, that, apart from the accepted differences in the landmarks, there was a significant variation in the weight of the female mandible. This weighed, on average, 79% of the male mandible whereas the differential for the female cranium is 86% of the male.

Mann (1989) devised the data shown in the Table 3.

Figures 3a, 3b and 3c illustrate the heavier and more defined landmarks in the male skull.

No gender can be assigned with accuracy to what have been considered to be juvenile skulls because differences do not appear to develop until puberty has been reached (Johnston *et al.,* 1958). The juvenile skull usually appears to be female. When all the permanent dentition has erupted and before evidence of wear occurs, a decision about gender may possibly be made depending on whether the spheno-occipital suture has closed *(vide infra).* Snow (1982) said that this is not a true suture but a synchondrotic joint and called it the spheno-occipital synchondrosis. This closure is of particular use when the skull appeared to be adolescent. If the suture is closed then the skull is adult and the gender is female, if open it is adolescent and the gender is still unresolved.

Changes occur in elderly skulls which make the gender difficult to interpret. When the dentition is lost, the individual's ability to chew fibrous foods is reduced. Because of the diminished use, the musculature reduces in bulk and the muscle attachments resorb and become less defined.

The condylar measurement can also be used to help to ascertain gender (Whittaker *et al.,* 1990b).

4.2.2 Gender from the dentition

There is a difference in eruption times between males and females as was indicated in 4.1, with the female dentition consistently erupting earlier. Hillson (1996) reported sexual dimorphism in crown size but that there was only a very small discrepancy of 0.4 - 0.5 mm and this is therefore likely to be a source of inter- and intra- observer error.

4.3 Age determination from the skull and the dentition

4.3.1 Age determination from the skull

In using the skull for age determination, the margin of error is greater when degenerative changes have occurred in the mature skull (Snow, 1982). Another source of error might be that there was a '*slower maturation in antiquity – as it is among modern Nubians, who reach puberty two to three years later than modern Americans.*' (Wente,1995).

In the infant, the anterior fontanelle (Last, 1981) is in the vault of the skull at the site of the bregma, where the two parietal bones and the frontal bone meet. By the age of about 18 months this fontanelle has ossified. Its presence allows the skull to mould in the pelvis during birth, and prevents damage to the brain. In the infant, the proportion of cranium to facial skeleton is large in comparison to the adult proportion. The mandible of the edentulous infant is similar in shape to the mandible of the edentulous elderly but is much smaller in anterior/posterior and bilateral dimensions.

As well as its use for gender identification (4.2.1) the spheno-occipital synchondrosis is also used as an aid to age determination. The synchondrosis closes in males at about 20 years and in females, rather later (Abbie, 1950). It is a reliable indicator of age unlike the other cranial sutures in which there is much variability in the time of closure (Miles, 1962). Suture closure in the young tends to originate at the anterior of the suture. In the adult and the male, suture closure changes to posterior closure (Abbie, 1950). Figure 4a shows an open synchondrosis with an unworn dentition and figure 4b is of a closed synchondrosis with the dentition worn.

The skull becomes thinner and lighter in old age. Osteoporosis is thought to be caused by a reduction in the formation in the bone matrix while the bone absorption continues as before (Soames & Southam, 1998, Boyd, 1958). Tooth loss results in the reduction of the size of the maxilla and mandible because of the resorption of the alveolar bone and there is also a reduction in the vertical height of the face. It has been observed that the mental foramen and mandibular canal are, therefore, closer to the alveolar margin. The angle of the ramus changes and becomes similar to the infant mandible in shape but, in the infant, the foramina are lower in the body of the bone.

4.3.2 Age determination (from the dentition).

Authentication of age by the dentition can be evaluated in several ways. The first assessment of age is made from the development of the dentition and the second by the degree of wear. In the first decade of life the estimate of age is within one year of the chronological age and in the second decade within two years.

4.3.2.1 Development of the dentition

The primary dentition starts to erupt at 6 months and is complete at about 3 years (Figure 5a). The permanent dentition commences eruption at about 6 years with the first permanent molars and the loss of the deciduous, lower incisors followed by the upper incisors (Figure 5b). The remaining primary dentition has usually exfoliated by the age of 12 years when the second permanent molars have erupted. The eruption of the third permanent molar is variable. The average estimate is at 18 + / – 3 years (Figure 5c). Hillson (1996) quoted Ubelaker (1987) who wrote that '*different methods might produce different age estimates from the same material*'. Ubelaker had revised the chart by Schour and M assler (1941) which is the standard used in anthropology.

4.3.2.2 Age changes in the mature dentition

Age determination in the deceased is difficult after the eruption of the third molars. Gustafson (1950) and Scott and Symons (1977) gave a list of age-related dental changes. These changes are:

- *Enamel and attrition.* Because of occlusal wear, there is flattening of the cusps from loss of enamel. This loss also occurs on the interstitial surfaces of the teeth and the total length of the dental arch is therefore shortened. The enamel tends to becomes susceptible to fracture with age because of changes to the dentine and pulp. *(vide infra)*
- *Secondary dentine formation.* There is deposition of regular secondary dentine throughout life. This slows down with age. The size of the pulp chamber is thereby reduced (Scott and Symons, 1977). When the dentine is exposed by attrition, irregular secondary dentine is formed by the odontoblasts lining the pulp chamber in an attempt to protect the pulp from exposure. With age, the dentine tubules of the root become calcified gradually spreading from the apex to the crown and this is an accurate method for assessing age. Begg [1] (1954) observed that secondary dentine is laid down even when there is neither wear nor caries and in old age the pulp chamber may be completely obliterated.
- *Pulp.* There is a reduction in the vascularity of the pulp with increasing age. In consequence the intact odontoblast layer lining the chamber is fragmented. The collagen content of the pulp increases and forms scar tissue. The dentine dries and shrinks. This is one of the causes of fracturing enamel in the dentition of the elderly and the brittle enamel found in archaeological specimens.
- *Apposition of cementum.* There is a continuous deposition of cementum at the apex throughout life which compensates for the loss of height of the crown because of the continuous wear of the tooth. Whittaker *et al.*, (1982) found that, in the mature dentition, the loss of crown height from attrition was of the same magnitude as the distance between the amelo-cemental junction to the alveolar crest.
- *Periodontal breakdown.* This causes the loosening of teeth, which Gustafson (1950) classified together with continuous eruption. These are different processes and should be considered separately.
    - **a.** As the occlusal surface is reduced in height because of wear, the tooth maintains the height of the clinical crown because of the deposition of the cementum at the apex.
    - **b.** The gingival margin migrates towards the root with age exposing the cementum.

a. and b. may be the same physiological process as these are features of continuous eruption.

- **c.** There is a loss of gingival attachment caused by periodontal disease.
- *Resorption of the root.* As this condition can only be confirmed on microscopic evaluation, it is not of value where a non-destructive examination is required.

4.3.2.3 Age assessment from 'attrition' (wear)

The most useful estimation of age of dental specimens after the age of 18 years is by a comparison of attrition rates, as discussed by Miles (1962). He considered that, in the young, the only reliable evidence of age is by observing which of the epiphyses is not yet united in the long bones or by determining if the dentition has not fully erupted. Miles (4.3.1) considered the cranial sutures and other bony landmarks to be very unreliable. He examined the rate of wear of the teeth. Comparing a group from within one community and arranging the skulls under the age of 18 years in order of eruption age, he noted that the wear on the teeth was consistent within the group. For example, when the second molar had just erupted, the first molar exhibited 6 years' wear. When the third molar had just erupted, the second molar showed 6 years' wear and the first molar, 12 years' wear. He then examined skulls with increasing wear on the third molar and, by this method, aged them.

Miles noted that the wear ratio of the wear on M1: M2: M3 teeth was 6: 6.5: 7. This means that it takes 6.5 years for the second molar and 7 years for the third molar to exhibit the same amount of wear as the first molar does in 6 years. For this study Miles had examined a group of Anglo Saxons from 700-800 AD. This was over 4000 years later than the predynastic Egyptians whose dentitions were the earliest examined in this survey. They were also a different ethnic group and the diet would have been very different. These factors might change the degree of wear but the ratio would remain unchanged.

Ageing a single dentition by the level of wear alone would not be valid, unless there were other examples of dentitions from the same milieu, because the local diet would be one of the main factors which might influence the wear.

Another example of classification of age by attrition can be made using a table (Figure 11) by Brothwell (1963). In this he had made a provisional classification of age in Neolithic to Medieval British skulls in three bands of severity. He has illustrated this in a table showing the gradient of wear expected between the 1st, 2nd and 3rd molar. He has shown three levels of wear on the 1st molar that could be expected between the ages of 17-25 years. For each of the three levels he has then added the anticipated wear for the next 10 year period until the age of 35, then 45 years. Any further wear is interpreted as 45+ years.

Teeth may be lost ante- or postmortem, may be unerupted, or may be congenitally absent. It is impossible, without radiological verification, to be certain whether teeth are unerupted or congenitally absent. As a general rule, an alveolus, containing an unerupted tooth, tends to appear expanded although this is not a *sine qua non*. The socket of a tooth, which has been lost postmortem, is usually easy to identify because there is no healing in the socket which, in health and in the presence of a tooth, is lined with dense compact bone, the lamina dura. When the tooth has been lost ante-mortem there is healing evident at the site of loss. The amount of healing is indicative of how long before death the tooth has been missing. The bone regenerating in the socket is cancellous bone and can thus be recognized macroscopically. The socket has usually disappeared 20-30 weeks after the loss of the tooth (Soames & Southam, 1998). Periodontal disease may make it difficult to determine how long before death the tooth was removed because the socket may be shallow.

4.4 Level of Wear (Attrition)

The gradient of attrition between adjacent molars has been discussed in section 4.3.

In the upper 1st molar the dentine is first exposed on the mesio-palatal cusp. This is followed by exposure on the mesio-buccal cusp, the disto-buccal cusp and finally, on the disto-palatal cusp. There is seldom any variation in this order (Murphy, 1959). Usually, as the wear proceeds, the two palatal exposures of dentine coalesce. The mesio-buccal aspect is then involved and finally, the disto-buccal exposure. Then there is a bridge of enamel left between the two buccal exposures of dentine. As this wears, a rim of enamel is left surrounding the exposed dentine. This is worn mesially, palatally, distally and then buccally. The wear is, therefore, greater in the palatal aspect of the occlusal surface of an upper molar.
In the lower 1st molar, the exposure of dentine is first seen on the mesio-buccal cusp. The disto-buccal cusp, then the mesio-lingual cusp and, finally, the disto-lingual cusp follow this. The dentine first coalesces at the buccal aspect, then the mesio-lingual and, finally, the disto-lingual. An area of enamel then surrounds the dentine which is lost first on the buccal aspect. When the enamel surround is lost on the palatal aspect of the upper molar and the buccal of the lower, as the dentine is softer than the enamel, the wear is oblique (Hillson, 1996)

The 2nd molars in both arches show a similar wear pattern.

The upper third molar loses enamel first on the mesio-buccal cusp, and the lower third molar on the mesio-lingual cusp. This results in two different occlusal planes. The slope in the first molar region is in a buccal direction and, in the third molar, in a palatal direction. This is known as a helicoidal plane.

The atypical wear noted in the molar of Piltdown Man was part of the evidence indicating that the remains were a forgery. In 1954, more than 40 years after the announcement at the Geological Society of London of its discovery, the specimen was reassessed. Not only were the teeth and jaw then recognised to be that of an orang-utan but the teeth had been artificially abraded (Encyclopaedia Britannica 2002).

A scoring system will be discussed later in this chapter.

4.5 Examination of the mandibular condyle

Various aspects of the mandibular condyle can be examined and measured.

4.5.1 The mesio-lateral measurement

The mesio-lateral measurement of the mandibular condyle can be registered using Vernier Calipers at the widest points of that dimension. Whittaker *et al.*, (1985) measured right and left condyles from 150 skulls of known gender and found no statistical difference between either side. He also found no statistical difference between male and female skulls in the antero-posterior dimension. The mean measurement of the mesio-lateral dimension of the male condyle was found to be 21mm and the female condyle, 18.5mm. These were Anglo Saxon skulls and there may be significant differences in the dimensions from the Egyptian skeletal group but the principle of the variation in size will probably be valid because of the increased musculature of the male skull.

4.5.2 Form and surface of the condyle

Changes in both the condylar form and surface integrity have been measured as an indication of pathological modification or osteoarthritis (Wedel *et al.,* 1978 and Richards & Brown, 1981).

Wedel recorded the observed changes as;

Form change

0. Without changes.
1. Slight remodelling.
2. Marked remodelling.
3. Deforming change.

Surface change

0. Without changes.
1. Uneven surface, intact, compact layer.
2. Irregular surface, perforation of compact layer.
3. Compact layer broken up >3mm².

Wedel then added the two scales to make *an index of form and surface changes* as an index of osteoarthritis of the temporomandibular condyle.

4.6 Continuous eruption

This is a physiological change which can be erroneously interpreted as periodontal disease. Because of the severe wear experienced in the dentition of the ancient Egyptian, the height of the anatomical crown was rapidly reduced. The anatomical crown is the part of the tooth from the amelo-cemental junction to the occlusal surface. The clinical crown is defined as the part of the tooth from the alveolar crest to the occlusal surface. It has been noted that there is a continuous eruption of the tooth throughout life to maintain the height of the clinical crown (Levers & Darling, 1983, Whittaker *et al.,* 1985 and Whittaker *et al.*, 1990a).

4.7 Angle's classification

Begg [2] (1954) stated that the '*teeth continually migrate throughout life in two directions: horizontal (mesial migration) and vertical (continuous eruption)'.* He also considered that the persistence of cusps in an adult was a prolongation of the juvenile form. Anatomically correct occlusion develops because of the horizontal and vertical movements and the alteration of the occlusal surface because of wear. He also stated, controversially, that there is a requirement that the primary and permanent dentition, before there is any wear, be greater than the alveolar bases. This is because Stone Age man would have *'insufficient tooth substance to occupy fully the tooth-bearing parts of his jaws because tooth attrition is so extensive'*

When both the maxilla and mandible are present, the Angle's classification can be made. It might be possible to demonstrate an ethnic relationship or even kinship within a group. When skeletal material is being examined the problem often found in wrapped mummies is eradicated as these teeth may not always be in occlusion because of the quantity of packing material in the oral cavity. Sometimes, the wadding material, impregnated with resin, is radio-opaque and obscures the dentition. Occasionally, the mandibular condyle is forced forwards or backwards out of articulation with the glenoid fossa during mummification, making it impossible to assess the relationship (Harris *et al.,* 1980). The assessment of the Angle's Classification can often be very difficult or impossible because of the postmortem loss of teeth. The landmarks used are: -

- The relationship of the 1st upper and 1st lower molar.
- The overjet, which is the relationship between the upper and lower incisors.
- The A-point is the deepest point on the incisive fossa of the maxilla.
- The B-point is the deepest point on the incisive fossa of the mandible.

To use the A- and B-points it would be necessary to have lateral cephalograms. These are not always available in a large study but may be used when an individual mummy is examined.

4.7.1 Angle's Class I

In an Angle's Class I relationship the A-point is almost vertically above the B-point. The mesio-buccal cusp of the upper 1st molar occludes in the median fossa of the lower 1st molar. The overjet is not more than about 3mm.

4.7.2 Angle's Class II

In an Angle's Class II relationship the A-point is anterior to the B-point. The disto-buccal cusp of the upper 1st molar is either vertically above the disto-buccal cusp of the lower or occludes in the median fossa of the lower. The overjet is increased and the maxilla is prognathic or the mandible is recessive. There are several examples of this in the New Kingdom Royal Mummies. Queen Tetisheri, the wife of the last Pharaoh of the 17th Dynasty, and her granddaughter, Queen Ahmose Nefetari, of the 18th Dynasty, both had prognathic maxillae (Harris & Wente, 1980).

4.7.3 Angle's Class III

In an Angle's Class III relationship the B-point is anterior to the A-point. The lower 1st molar is mesial to the upper 1st molar and there is a reverse overjet. The lower anterior incisors occlude anteriorly to the upper incisors. The mandible is prognathic. Begg & Kesling (1977) considered that '*attritional occlusion*' resulted in a Class III relationship.

4.8 Additional information

Additional information may be noted. These are:

- Anomalies such as supernumerary, malformed or congenitally absent teeth.
- Whether the brain was removed and which pathway was selected. Brain removal was practised from the Middle Kingdom but became common practice during the New Kingdom. The usual method was through the cribriform plate of the ethmoid bone via the left nostril. It is also documented that King Ahmose, the first Pharaoh of the New Kingdom, had his brain removed through the foramen magnum (Russell, Storey & Ponitz, 1980) and there is some record of excerebration through the orbit.
- In some osteological specimens a mandible and maxilla may be mismatched by gender and age (Figure 14).

4.9 Mummified specimens

These are much more difficult to assess than osteological specimens as they may be wrapped and then the only method of visualizing the dentition is by radiology. Even if the mummy has been unwrapped, the tissue is so rigid that usually the dentition can only be examined by radiology. The only exception to this has been Natsef Amun, the 'Leeds Mummy'. He had been mummified with his mouth wide open and his teeth partially visible (Miller & McDade, 1992).

4.9.1 Technique of mummification

When a body was buried in the hot sand, it desiccated and this prevented putrefaction of the body. This is known as natural mummification. These bodies remain in good condition even 5000 years later and, arguably, were the most successful examples of mummification. An example of this is a predynastic mummy, 'Ginger' who died around 3200 BC and was found in a pit grave in the desert. He is now displayed in the British Museum. The problem was that these bodies, at the edge of the desert, were often disinterred by wild animals and destroyed. The risk to the deceased was also increased by the practice of burying the body with grave goods which were to be of service in the afterlife. This custom encouraged tomb robbers. A better protection of the body was sought and elaborate tombs were constructed.

Known today as mastabas, these had a substructure cut into rock and were divided into rooms by mud bricks. The central room was the burial chamber and the surrounding rooms were used to house the grave goods. The superstructure was a mud brick rectangle up to 200 feet long and 80 feet wide which was also divided into storage rooms. There are no intact roofs in the remaining mastabas, but it is suggested that they were rounded with a flat parapet at each end (Watson, 1987). In rare instances, servants were buried in graves outside the surrounding walls, perhaps killed to accompany the pharaoh on his journey to the afterlife. At a later stage, sets of model representations, the ***ushabti*** figures, were buried in the tombs instead of sacrificing the living attendants. These miniature figures which appeared in the 9th Dynasty are mummiform in shape and carry agricultural tools which were painted or etched on the surface with the engraved name of the tomb owner (David, 1982).

The Step Pyramid appears to have been constructed by adding successive, but diminishing, layers of superstructure to a mastaba. This pyramid was designed in the 3rd Dynasty for King Djoser by his vizier, Imhotep, who was a magician-physician, sage and architect. By the Persian Period in 525 BC he had been deified to become the god of medicine (Hurry, 2000).

The security offered by building mastaba tombs and pyramids had significant adverse effects. Because there was such protection for the body, more valuable grave goods were provided to accompany it into the afterlife to ensure the well-being of its ***K3*** - the spirit or life force. Thus, temptation to tomb robbers became greater. However, the most devastating consequence to the body was that, because the deceased was no longer buried in proximity to the hot sand, there was rapid decomposition. It was necessary to find a way to preserve the body and so the first tentative steps towards artificial mummification were taken.

The bodies were first bandaged and coated in resin to preserve the shape but this did not prevent the internal decomposition. In the early Old Kingdom embalmers began to remove the viscera and there are canopic containers from this period. One example, from the 4th Dynasty, is a casket of alabaster divided into four which contains the viscera of Queen Hetepheres. The most successful innovation was the introduction of natron. This is a naturally occurring mixture of sodium bicarbonate,

carbonate, sulphate and common salt which dehydrates tissue.

The body was taken to the embalming house which was called ***w3bt*** the Pure Place, or ***pr nfr*** the Beautiful House. The abdomen of the cadaver which had been placed on an embalming table was marked by a scribe and the first cut with an obsidian stone knife was made by the ***paraschistes*** or 'ripper up' (Dunand & Lichtenberg, 1991). After evisceration and washing with palm wine and spices the abdominal cavity was filled with natron which then was also packed over the body. After 40 days the abdominal filling was removed and the abdomen was repacked with sawdust and resin soaked linen. The abdomen was then closed, only rarely being sutured. The surface was covered with molten resin.

Preservation was initially only performed on the pharaohs and the nobility. This means that there are many skeletal remains since the majority of the dead had neither been naturally nor artificially mummified. Gradually, as mummification became more sophisticated, it became more available to the population and by 1000 BC it was widespread. However the increasing numbers brought a decline in standards (Dunand & Lichtenberg, 1991).

Some of the skeletal material, which is available in such abundance in the museum collections, was used in the study reported in this thesis.

4.9.2 Examination of mummified specimens

Intact, wrapped mummies are difficult to examine. Many are in wooden coffins and are now seldom ever removed from them. This is because their condition is extremely delicate. Often the bandaging of the mummy has been stuck to the inner coffin by the resin which was poured over them. It is now considered unethical to interfere with human remains in a destructive manner.

It was not always so. In the 19th Century, travellers brought mummies home to England as souvenirs of their sojourn in Egypt. Many were unwrapped as an entertainment for guests who had been invited for the spectacle. Lord Londesborough had an 'At Home' on Monday, 10th June 1850 to *unroll* a mummy from Thebes (Figure 6).

Sir Percival Marling brought home the mummy of a child in 1886 which he gave to his father as a present. It stood in its coffin in his father's ballroom and was later bought in a sale by a Miss Phipps. This was given to the Manchester Museum in 1979 but had been left undisturbed until then (David, 1984).

However, there were several early investigations which were scientific and were not solely performed to search for amulets and other precious artefacts which were often included in the wrappings. In 1825, Dr. Granville examined the body of a woman who, he judged, was about fifty years old at death. He found that she had a cyst of the uterus and had probably died from pneumonia (Filer, 1995).
In 1828, the Leeds Philosophical Society examined the mummy of Natsef Amun, who had been a priest at the temple in Karnak. He lived during the New Kingdom, in the early 21st Dynasty (David, 1992). The results of this examination were published the same year with a full report of the unwrapping, the materials used, a chemical assessment of these materials and an anatomical examination of the mummy. In 1989, the Director of Leeds City Museum invited the Manchester Mummy Team, under the leadership of Dr. Rosalie David, to undertake a new multidisciplinary investigation.

n 1908 Dr. Margaret Murray, a former student of Sir Flinders Petrie and first curator of Egyptology at Manchester Museum, examined the two brothers, Khnum Nakht and Nakht Ankh, using a multidisciplinary approach. By that date she felt, '*every vestige of ancient remains must be carefully studied and recorded without sentimentality and without fear of the outcry of the ignorant*' (David, 1992).

Because of the almost non-destructive techniques of obtaining information using radiography and endoscopy, there is now no necessity to unwrap the mummy which therefore remains intact. These techniques are expensive and cost may limit their use. There are several different types of radiographs which are useful for a dental examination.

4.10 Radiographic examinations

4.10.1 Cephalograms

An initial radiograph must be taken to confirm the alignment of the mummy in the bandaging and cartonnage. The mummy might have been positioned with its head flexed or at an angle. The alignment can also be determined very successfully with a CT scan. When the orientation of the skull in its wrappings is established, then specific cephalograms can be taken. In the live patient, these radiographs are usually taken with rods inserted in the external auditory meatus on both sides to orientate the head so that the beam can be aligned at 90° to the cassette.

These radiographs are used to assess the craniofacial morphology and can then be employed as a means of comparison of possible kinship or ethnic features when there are several mummies from the same archaeological site using fixed object to film distance for comparative measurements. It can also demonstrate the path by which the cranium has been excerebrated as it will show the fracture and displacement of the *sella turcica* if this has occurred through the ethmoid bone. However, it is essential to determine whether the condylar head is in the glenoid fossa before interpreting the radiograph for jaw relationships. The temporomandibular joint is often disturbed during mummification. This radiograph also records the teeth that have been lost and indicates the wear and oral pathology. There are problems in the accurate assessment of individual teeth as there is superimposition of the right and left quadrants.

4.10.2 Orthopantomograms

If the head is detached, an orthopantomogram might be the radiograph of choice. It cannot be taken when the head is horizontal as with an intact mummy because the machinery is large and has been designed for the standing patient. This x-ray gives a panoramic view of the complete dental arch. When the head is that of a young person, it is possible to estimate the age from the stage of eruption of the permanent dentition. Pathology, in the dentition and alveolar bone of the maxilla and mandible, is seen easily. Sometimes it is possible to observe the condition of the mandibular condyle. It shows periapical pathology at an earlier stage than it is possible to detect from a direct inspection of a skull where it may be overlooked.

4.10.3 Intraoral radiographs

When Natsef Amun, the Leeds mummy, was examined in Manchester in 1991, he had been unwrapped previously in 1828 at the Leeds Philosophical Society. It was possible to attempt a clinical examination as, unusually, he had his mouth wide open with his tongue protruding (Figure 7a) as reported by Miller & McDade, 1992. Intraoral radiographs were inserted and secured with 'Blue-Tack' (Bostick Ltd.). This may, as yet, be the only mummy which has been investigated using intraoral films. These films (Figure 7b) give excellent detail of groups of teeth and their supporting bone.
Another example of an intraoral film is the occlusal film which was used to good effect when the skeleton of Nakht Ankh, one of the two brothers, was examined by Leek (1979). It showed the large geminated central incisors and a supernumerary lateral tooth.

4.10.4 Computed tomography.

Computed tomography is now regularly used as this has the advantage of causing no disturbance to the mummy which can be left in its coffin. The body is moved through the beam and digital information is registered on the Hounsfield scale. The range is from +1000 for bone to –1000 for air. The image is reconstituted as an analogue. Slices are taken horizontally, saggitally and coronally and images can be seen at selected planes. The information may also be reformatted to give a three dimensional impression of the anatomy. Digital information, so obtained, can also be used to tool a skull in polystyrene. Another technique is Stereolithography which uses an argon / ion laser to polymerise acrylic resin models from the digital data. Both of these make a base for facial reconstruction. Unfortunately C.T. scans of the dentition are sometimes very difficult to interpret because of skull orientation or orientation of the slices.

4.11 Dental pathology found in Ancient Egyptians

Although the dental condition from which the Egyptians most suffered has been labelled *attrition*, the problem was not caused by attrition alone. Because of the contamination of the bread by quartz, the effect was rapidly augmented by abrasion (2.7.2.). The distinction between attrition and abrasion is not as clearly defined as it appears in the following sections. It is realised that there is an element of both processes in each condition. A better description of the process is given by Soames and Southam (1993) 'as *tooth wear with a major component of attrition / abrasion.'*

4.11.1 Attrition

Attrition is defined as '*the physiological wearing of teeth as a result of mastication'*, (Shafer *et al.*, 1958[1]) or '*loss of tooth substance as a result of tooth to tooth contact'* (Soames and Southam, 1998). It occurs on the occlusal and interproximal surfaces. The occlusal surface looses enamel and the cusps flatten. When the attrition is sufficient to expose the dentine, the odontoblasts are stimulated to form secondary dentine in the pulp chamber to protect the pulp from exposure and its subsequent death. The interproximal wear appears at the contact points of adjacent teeth because of their vertical movement within the periodontal membrane, during mastication. The contact between adjacent teeth progresses from a contact point to a complete surface of contact. Attrition, *per se*, is not pathological wear. It is found in groups of people who have a fibrous diet (Begg and Kesling, 1977 and Hinton, 1982). Although this partly explains the severe wear in the dentition of the ancient Egyptians, it does not completely clarify the aetiology of the pathology which has been found frequently. In addition to the fibrous diet there are quartz phytoliths, included in the cellulose structure, which have an abrasive effect.

There are several scoring systems used to assess attrition but the scheme illustrated by Molnar (1972) is the most detailed (*vide infra* 5.6.1). Because of Hodges' assertion (1991) that there was a 24% error in a blind re-examination of attrition using Molnar's scale, Miller (1997) measured the attrition in 75 mandibles using both Molnar's scale and Smith's (1979), less specific, scale. The percentage error found was 17% when the results with a discrepancy were re-checked. In general it was found less problematic to classify the wear using Molnar's scale. As this is the scoring system which has been used in this study there is a detailed account in 'Materials and Methods', Chapter 5.

4.11.2 Abrasion

Abrasion is defined as '*the pathologic wearing away of*

tooth substance by the friction of a foreign body independent of occlusion.' (Soames and Southam, 1998). Most of the papers about the extreme wear in the dentition in the ancient Egyptian identify the wear as 'attrition' (Leek, 1973, Leek, 1986 and Grilletto, 1969). However, severe attrition on its own is unlikely to have caused the devastation seen in the dentitions. Begg [3] (1954), in his description of the attrition in the Stone Age Australian Aboriginal, in which the dentition was frequently worn down so that the roots had erupted into the occlusal plane, illustrates this in his paper showing the recession of the pulp chamber with no exposure of the pulp. The Egyptian often experienced the exposure of the pulp because the laying down of the secondary dentine was overtaken by the rapid wear of the dentine. This is an example of '*tooth wear with a major component of abrasion*'.

4.11.3 Erosion

Erosion is defined as '*the loss of tooth substance by a chemical process that does not involve known bacterial action*' (Soames and Southam, 1998). The pH of the saliva is reduced. It may be caused by drinking acidic drinks such as fruit juices (Figure 8a and 8b). Other causes may be frequent vomiting because of illness, gastric reflux or bulimia nervosa. The incisors of Natsef Amun appeared to have been affected by erosion.

4.12 Caries

'*Caries is a disease process characterised by the focal demineralisation of dental hard tissues by organic acids produced by bacterial fermentation of dietary carbohydrates, especially sugars*' (Larsen *et al.*, 1991). The effect of the disease is the development of a cavity which might result in necrosis of the pulp. Larsen and his co-worker's dental sample was taken from the south east coast of the USA from 1000BC – 1150AD, which they defined as the pre-agricultural and pre-contact group and from 1150AD – 1550AD, the agricultural, but still pre-contact group. There were 460 individuals in these two groups. They found that there was a higher rate of caries in the females and gave two reasons for this. The first explanation was that the females' dentition erupts earlier than that of the males, and secondly, females cared for crops and the preparation of food when the males hunted. Hillson (1996) defined dental caries as '*the destruction of enamel, dentine and cement resulting from acid production by bacteria in the dental plaque, ultimately leading to the formation of a cavity in the crown or root surface*'. This was first recognised by Miller in 1889 (Soames and Southam, 1998) and the definition is still accepted. The dental plaque contains bacteria of various kinds and *Strep. mutans* is most commonly associated with caries. However there is sometimes a high concentration of *Strep. mutans* when there is no caries and *vice versa.* Dietary sugars are converted to acid by the bacteria, which demineralises the enamel and this leads to the destruction of the organic matrix. Sucrose, a disaccharide containing fructose and glucose, is the most cariogenic of the sugars. The individual monosaccharides are also highly cariogenic. Those most commonly implicated are glucose, fructose and maltose.

It has been found that those who suffer from fructose intolerance, one of the monosaccharides contained in sucrose, but who can tolerate polysaccharides, do not suffer from caries. It is not the total amount of sugar consumed that is the problem but rather the frequency with which it is taken.

Dietary sugars have been classified as intrinsic and extrinsic sugars (Edgar, 1993). '*Intrinsic sugars are integrated into the cellular structure of the food. Extrinsic sugars are free in the food or, added*'. Extrinsic sugars, which are added to food, cause more destruction than intrinsic sugars contained in food such as fruit and vegetables. (Soames and Southam, 1998). Fruit and vegetables have intrinsic sugars but fruit juice, milk and honey have extrinsic sugars. Lactose, a monosaccharide which is found in milk, is less cariogenic than the other monosaccharides. However fruit juices, which contain extrinsic sugars caused by the destruction of the cellular structure, are responsible for causing caries. Despite the fact that fruit in the diet contains mainly intrinsic sugars, it lowers the pH of the saliva and therefore could also be responsible for some increase in caries.

At the point that disaccharides in fruit juices and honey, which were used as sweetening agents, were introduced to the diet, caries began to appear in childhood and attacked both of the most frequent sites for caries, the fissures and approximal surfaces. Prior to this time, because of the rapid attrition, the occlusal surface was flattened and the fissures were obliterated before the dentine was penetrated. Also, because there was approximal attrition, the ancient Egyptian suffered less from approximal caries. Other surfaces attacked are the interproximal and the root surface at the amelo-cementum junction. There is significance in the site of caries. Grilletto (1973) found that root caries was most common in the Neolithic Period.

4.13 Periapical lesions

Soluble toxins from bacteria, which are present in carious lesions, disperse through the dentinal tubules and cause the inflammation of the pulp even before the secondary dentine barrier has been breached by caries. When the pulp chamber has been penetrated by attrition, it is specifically the bacteria that cause the pulpitis. The blood vessels dilate and inflammatory oedema develops. This compresses the circulation and the pulp becomes necrotic. Suppuration occurs with the breakdown products of leucocytes and bacteria. The pus is forced through the apex of the tooth and causes a periodontitis. If the suppuration continues, the abscess will enlarge and tracks through the cancellous, alveolar bone and eventually penetrates the dense cortical bone. In the ancient Egyptian skulls, this is what is seen as well-defined, circular holes at the apex of the roots, and which has been previously described as conclusive proof that these were made by the ancient Egyptian, with fire drills, to relieve the pressure of the abscess (Figure 9).

4.14 Periodontal changes

It is necessary to distinguish between the periodontal changes that occur as the result of disease from the physiological changes caused by ageing. This condition is initiated by gingivitis, which is the consequence of the microbial plaque around the neck of teeth. As nothing is known about the oral hygiene habits of the ancient Egyptians, except for their use of natron masticatories *(vide supra 2.5.5)* for purification of the mouth for religious and social reasons, it is likely that the plaque around the cervical margin might have adhered, undisturbed, for a considerable period of time. First, there would have been acute inflammatory changes, such as vasodilatation, resulting in the exudation of fluid in the crevice between the gingival tissue and tooth. This fluid contains polymorphs and immunoglobulins, which would initially help to control the effects of the gingival infection (Soames & Southam.1998). Eventually, it would result in the destruction of the attachment of the gingiva to the enamel. This would lead to the deposition of subgingival plaque and further apical movement of the attachment and loss of the supporting bone.

There are two kinds of periodontal bone loss, which are noted in the skulls, horizontal and vertical.

4.14.1 Horizontal bone loss

In the initial stages this bone loss may be easily confused with the apparent horizontal bone loss seen in the skulls, which is the result of the continuous eruption of the tooth, which is in part the physiological response to the rapid attrition. Generalised periodontal disease, initiated by plaque retention, is eventually responsible for the loss of most of the buccal wall of the alveolar bone and the tooth may become so mobile that it exfoliates. In the early stages it can be recognised by pitting in the compact bone of the alveolar crest (Figure 28a).

4.14.2 Vertical bone loss

This is very different in appearance and is the result of an acute periodontal infection, which has resorbed the alveolar bone unevenly.
Because of the condition of the mummified tissue it is impossible to evaluate the soft tissue. However, because of large collections of skeletal remains, it is possible to assess the skeletal destruction caused by periodontal lesions.

4.14.3 Physiological bone loss

This is the apparent horizontal loss of alveolar bone, which may be measured as the distance between the amelo-dentinal junction and the alveolar crest. This may be the result of the continuous eruption of the tooth to compensate for the attrition and is not the result of a pathological process (Kerr & Ringrose, 1998).

While the aetiology of periodontal disease relates to the presence of dental plaque, there are other factors which influence the progress of disease such as nutritional deficiencies and systemic disease (Soames & Southam, 1998). In ancient Egypt there were nutritional variations caused by drought and flooding. There is historical evidence for this in the literature and skeletal evidence of arrested bone development, the Harris lines, has been found in the radiology of the long bones. Nutritional disturbances and childhood illnesses also cause hypoplasia in the enamel. There is also evidence of diseases such as diabetes mellitus, and blood dyscrasias such as anaemia (Estes, 1993), which cause gingival inflammation and could therefore result in bone loss and periodontal destruction.

4.15 Cysts

A cyst is a pathological cavity having fluid or semi-fluid contents which has not been created by the accumulation of pus (Soames and Southam 1998).

They are more common in the jaws than in other bones and may be of dental or non- dental origin.

4.15.1 Radicular cysts

The most common cystic lesion is associated with a non-vital tooth. The tooth is often symptom free. The cyst is related to a granuloma, which occurs when the infection is chronic rather than to pus which is the result of acute infection. There is expansion of the bone caused by the deposition of new bone by the periosteum covering the cortical bone. The bone surrounding the granuloma is resorbed as the cystic area enlarges. As the expansion is faster than the deposition of bone the cortical bone thins. The cystic area crackles like an eggshell when palpated during life. Post mortem, the bone is so thin and delicate that it usually fractures. Cysts are more common in the maxilla than the mandible and may be the result of aseptic pulp death following trauma.

These are the most prevalent dental cysts.

4.15.2 Dentigerous cysts

These are associated with teeth which are impacted or are late to erupt. They are twice as common in males as females and twice as common in the mandible as the maxilla. The most usual site for dentigerous cyst formation is the mandibular third molar. The next most frequent is the maxillary canine.

4.15.3 Nasopalatine cysts

The nasopalatine cyst is a developmental cyst which forms from the embryological, epidermal residue in the duct connecting the oral and nasal canals.

4.15.4 Rare bone cysts

There are some unusual cysts such as the odontogenic keratocyst. This evolves from vestiges of the dental lamina. Stafne's idiopathic bone cavity is a developmental anomaly found posterior to the premolars and below the inferior dental canal and contains salivary tissue. The ameloblastoma is a locally invasive tumour which may appear similar to a odontogenic cyst (Soames & Southam, 1998).

4.16 Osteoarthritis of the temporomandibular joint (TMJ.)

As this is a degenerative disease of the joint and is non-inflammatory, it should be regarded as osteoarthrosis. The TMJ is unusual because the articular surface is covered with fibrous tissue rather than the hyaline cartilage of the other joints. Osteoarthritis is a recurrent finding in ancient Egyptian condyles and there have been various theories about its aetiology. Seward (1976), in his research into the temporomandibular joints of the Australian Aborigines, concluded that there was a relationship between attrition and wear of the joint. Richards and Brown (1981) agreed with this finding. However Eversole *et al.,* (1985), who examined a 20th Century skull collection, found no significant relationship, nor did Whittaker *et al.,* (1985) who examined the mandibular condyles of a Romano-British population. Miller (1997) suggested that, having found no significant relationship between attrition and osteoarthritis in ancient Egyptian skulls, there might be a third factor involved - the loss of facial height. This will be investigated in this study.

4.17 Osteomyelitis

Osteomyelitis of the jaw is a not infrequent finding in the ancient Egyptian skulls. The most likely cause was acute infection such as a dental abscess following exposure of the pulp caused by attrition or, less frequently, by caries. The other major causes might have been fractures or penetrating wounds to the jaw. The infecting bacteria multiply in the cancellous bone and the suppuration results in necrosis of the bone. Sometimes the necrotic bone is surrounded by pus and the bone forms a sequestrum. As there does not appear to have been any attempt to extract teeth, it is possible that many sufferers developed septicaemia which would have caused death.

4.18 Dental Anomalies

Several genetic anomalies in dental development have been found. There may be more to be discovered.

4.18.1 Hypodontia, supernumerary teeth and rudimentary teeth

Anodontia is the total absence of teeth and hypodontia or partial anodontia is partial absence of teeth. There do not appear to be any examples of anodontia in the examined Egyptian skulls. Hypodontia is less common in the primary dentition than in the permanent dentition. It is often bilateral and most often affects the third molars and, to a lesser extent, the lateral incisors and the mandibular second premolars. Brothwell *et al.,* in 1963, said that, because the absence of teeth other than 3rd molars or lateral incisors is infrequent, only these teeth have been studied in anthropological reports. One of the pair of affected teeth may be present in a rudimentary form. The upper lateral is occasionally 'peg-shaped' and the third molar can be very small and have an unusual morphology. Examples of these have been found. Lower incisors are even less commonly absent than those previously mentioned (Soames & Southam, 1998) but Mummy No. 10881, in the Manchester Museum, was thought (Leek, 1979) to have congenitally absent lower incisors. Supernumerary teeth are also occasionally found. These are often smaller than the rest of the dentition and are of an unusual shape. If the morphology is normal, the supernumerary tooth is called a supplemental tooth.

4.18.2 Gemination or Fusion

There are two conditions which are almost impossible to differentiate. (Soames & Southam, 1998). These are *gemination* and *fusion.* Gemination is the partial development of two teeth from a single tooth bud following incomplete division. Fusion is defined as the union between dentine and / or enamel of two separate developing teeth. The condition may be bilateral.

This anomaly was found in Khnum Nakht (Leek, 1979), one of the two brothers from the Middle Kingdom, in the Manchester Museum collection. His two very large central incisors are examples of partial and complete gemination. The right upper incisor is a complete gemination; the pulp chamber and canal are as one and the only notable feature is that the tooth is larger than anticipated. The left incisor is even larger and is an incomplete gemination. The enamel and dentine are united but the pulp chamber and canal are separate.

4.18.3 Hypercementosis

This condition is caused by the deposition of abnormal amounts of cementum. Usually the thickness of cementum increases with age and compensates for occlusal wear. Very occasionally, there is an excessive mass of cementum laid down in response to periapical inflammation (Soames & Southam, 1998) or heavy occlusal wear (Brothwell, 1963). Examples have been found in Neanderthal man.

4.18.4 Taurodontism

A taurodont tooth is one in which there is no narrowing at the cervical margin and the pulp chamber extends into the root. It may be found in conditions such as Klinefelter's syndrome (Soames & Southam, 1998).

4.19 Alignment of arches and jaw relationships

The alignment of the ancient Egyptian's dental arches is usually excellent. Because of the interstitial attrition, the mesiodistal widths of the teeth are reduced, and there is much less crowding than is found in the modern dentition where the diet is much less fibrous and abrasive. Of the 21 dentitions that Leek examined from the Manchester collection only three showed crowding. There was a child and adult from the same grave which may have indicated a close relationship. The third dentition was from a neighbouring grave which may also affirm a family connection. There were two skulls, which had mesio-angular impactions of the lower third molars, and one skull in which the supporting bone was pitted which may be the sign of recurrent infection around an erupting third molar (Leek, 1986).

The occlusion has been much more difficult to assess. Many of the mummies have their teeth out of occlusion because their mouths have been packed with linen and resin during mummification. Sometimes the mandible has been pushed backwards or pulled out of articulation during mummification, (Harris *et al.,* 1980) resulting in the false appearance of a Class 11 or Class 111 occlusal relationship. The majority of skulls appear to have a Class 1 relationship. Some major exceptions have been the occlusions of Queen Tetisheri of the 17th Dynasty, and her grand daughter, Queen Ahmose-Nefetari of the 18th Dynasty (Harris *et al.,* 1980[2]) who have severe Class 11 relationships.

4.20 Aims and objectives of the studies reported in this thesis

The examinations that have been previously carried out on skeletal and mummified material have given rise to the following speculations which have been investigated.

4.20.1 The Null Hypotheses for this study

1. There is no significant difference in gender and age at death between samples from each time period.
2. There is no significant difference in antemortem tooth loss between samples from each time period.
3. There is no significant difference in the wear on the molars between samples from each time period.
4. There is no significant difference between the condylar size and assessed genders.
5. There is no significant difference in degree of osteoarthritis between samples from each time period.
6. There is no significant difference in caries rate between samples from each time period.
7. There is no significant difference in the periodontal breakdown between samples from each time period.
8. There is no significant difference in the numbers of abscesses and cysts between samples from each time period.
9. There is no direct relationship between ratio of continuous eruption and crown height and increasing age, wear in the molars and osteoarthritis of the temporomandibular condyle between samples from each time period.
10. There is no direct relationship between wear of the occlusal surface of mandibular molars and osteoarthritis of the temporomandibular condyle.
11. There is no significant difference in the anomalies found at different time periods.
12. There is no significant difference between the percentages of ethmoidal excerebration from the New Kingdom and later.
13. There is no evidence, using hair samples examined by mass spectrometry, of malnutrition which might be the cause of severe periodontal breakdown.

Chapter 5

Materials and Methods

5.1 Source material

The two most comprehensive collections of Ancient Egypt skeletal remains in the UK, at Cambridge University and the British Museum (Natural History), were accessed for this study.

The photographs of specimens for this thesis were taken with a Fuji Mx700 Digital Camera at a resolution which varied from 650,000 pixels to over 1 million pixels.

5.1.1 The Duckworth Collection

The Duckworth Osteological Collection is accommodated in the Department of Biological Anthropology in Cambridge. It encompasses several collections, two of which are of particular relevance for the research reported in this thesis. These are the Thurnam Collection from the Department of Anatomy and the Pearson Collection from London. These contain the Egyptological material. The collection is used for research purposes only and is not on display (Foley, 1999). The Egyptian skeletal material has been garnered from sites from all over Egypt and Nubia, and there are specimens from the Predynastic to the Ptolemaic Period, a time span of over 4000 years.

5.1.2 The Natural History Collection

The collection, in the Palaeontology Department of the British Museum (Natural History), is a compilation of several smaller collections. These are from the Royal College of Surgeons of England, Sir Flinders Petrie's Collection from the Petrie Museum, University College, London, the Oxford University Egyptological Collection and the Williamson Egyptological Collection which is also from Oxford. All of the material used in this study came from the Petrie Collection from Hawara.

5.2 Data collection

The collections in the School of Biological Anthropology in Cambridge and the Natural History Museum (British Museum) together contain over 5000 Egyptian skulls. These skulls, from a specific time period at identified sites, are individually boxed and numbered and are in no particular order either by age or gender. It was decided to examine as many skulls as possible from each time period (*vide infra* 5.2.1 and 5.2.2). Lists were supplied for each group from the collection in Cambridge with a code indicating whether the skull was 'E' entire, 'C' cranium only or 'F' fragmentary. The boxes containing the entire skulls were selected and were opened in sequence. However, some of these were disarticulated and so were then excluded. Others, which were discarded, had experienced postmortem damage to the cranium and dentition or fractures to the body of the mandible or temporomandibular condyles. One skull had had teeth repositioned and glued into incorrect positions and therefore had to be discounted. Those samples chosen for examination were in a satisfactory condition with little postmortem damage apart from tooth loss. If a dentition could be articulated, even though there were fractures in the condyles of the mandible, its inclusion was reconsidered.

When there was a shortage of skulls in an acceptable state, intact mandibles, without crania, were selected on the condition that the condyles were unbroken because it was possible to take most of the required measurements from these. If there were still insufficient numbers, a different burial site from the same period was then studied.

5.2.1 Sample size

It was assumed that it would be possible to examine at least 60 skulls from each time period. A larger number was examined if available.

5.2.2 Changes in prevalence with time

It was anticipated that the prevalence of severe tooth wear would decrease gradually with time from 65% in the Predynastic Period to 40% in the Graeco-Roman period because of the changes in the quality of the diet. Assuming the sample size in each of the 6 period groups is 60, an $\alpha = 0.050$ two-sided chi-square test of trend in proportions based on the logistic model will have a 90% power to detect a difference in proportions characterised by a trend of -0.21 and a variance term of 0.24 with an average proportion of 53% of severe tooth wear (Elashhoff, 2000).

5.3 Examination of crania

A careful inspection of the crania was initiated before the dental condition was considered. The crania in the Duckworth Collection and the Natural History Museum (British Museum) were examined in groups defined by period. Figure 10 is a map of some of the sites.
These were:

5.31 Predynastic Period (5000 –3100 BC)

The first group of skulls was discovered in excavations at El-Badari, Naqada, Hierakonpolis and Nubia and were all from the period before 3100 BC.

Considered to be in suitable condition for this research were 40 Badarian skulls, 40 from Naqada and 11 from Nubia and Hierakonpolis. Although the Nubian collection was vast, containing over 800 skulls, the majority were in very poor condition. A copy of the historic, hand-written excavation list from the First Archaeological Survey of Nubia in 1907-8 was very difficult to interpret. Because only 6 of the 43 excavated cemeteries contained predynastic skulls and had also been used for burials from all of the time periods, the identification of these skulls was problematic. The Hierakonpolis skulls were scattered within the collection and were also difficult to find.

5.3.2 Archaic and Old Kingdom Periods (3100 – 2181 BC)

The Archaic Period material (3100 – 2686 BC) contains 50 skulls from Abydos and 30 of these which were in a suitable condition were considered along with the Old Kingdom skulls. From the Old Kingdom (2686 –2181 BC), there are 89 skulls from Aswan, 84 from Dishasha, 331 from Tarkhan and 101 from the Pearson Collection which had been excavated during the First Archaeological Survey of Nubia and were considered to be Old Kingdom. All 6th Dynasty Old Kingdom skulls from Qau were examined. In total 120 skulls were selected for the survey. These were the 30 skulls from Abydos, 20 from Qau, 10 from Aswan and Dishasha, 30 from Tarkhan and 30 from Nubia (Pearson collection). Many of the others had missing mandibles or were in a fragmentary state.

5.3.3 Middle Kingdom (1991 –1786 BC)

In the Middle Kingdom collection, there were 300 skulls from Kerma in the Sudan and 54 from Qurna el Tarif. 70 skulls from Kerma and 20 skulls from Qurna el Tarif were examined.

5.3.4 New Kingdom (1567-1085 BC)

The New Kingdom has specimens from the Ramesseum, Qurna, Memphis, Saqqara and Gurob.
From this period, 50 skulls from sites in Qurna, 34 from Memphis and Gurob and 5 from Saqqara were selected as in suitable condition.

5.3.5 Late Period (664 –332 BC)

From the Late Period 60 skulls from Gizeh were examined. There are 120 skulls which were complete out of the collection which contained 1700 skulls. Even these crania were, on occasion, incorrectly matched with mandibles which were of the wrong age and gender. Where these were discovered they have been recorded.

5.3.6 Graeco-Roman Period (332 BC – 400 AD)

The Graeco-Roman Period is represented by 113 skulls from Dendera and some of the 800 from Nubia.

As many of these skulls are in very poor condition only 17 were examined and a further selection, housed in the palaeontology department of the British Museum (Natural History), was included. These were from the Roman period from Hawara at the edge of the Fayum in Lower Egypt. This site had been excavated by Flinders Petrie in 1911. 58 of these were in excellent condition and so 75 Graeco-Roman skulls were examined in total.

5.3.7 British Neolithic skulls

14 Neolithic skulls from the British Isles were included as a comparison with the Predynastic Group of Egyptian skulls.

5.4 Specific parameters evaluated were:

1. Age and gender at the various sites.
2. Ante mortem loss of teeth.
3. Wear on 1st, 2nd and 3rd molars.
4. Size of mandibular condyles as corroboration for assessed gender.
5. The extent of osteoarthritis of the condyle using a scale for evaluating damage to both form and surface.
6. The rate and distribution of caries.
7. Horizontal and vertical periodontal breakdown.
8. Damage ascribed to abscesses and cysts
9. Continuous eruption and the effect on its rate of development by age, molar wear and osteoarthritis of the mandibular condyle.
10. Anomalies
11. Dental relationship of the maxillary and mandiblular teeth.
12. Ethmoidal excerebration.

These are considered for each time period.

5.4.1 Gender

The gender of the skulls was assessed from the anatomical landmarks. Some of the distinguishing features between the genders are the heavier supraorbital ridge, the shape and contours of the orbit, the pronounced muscle attachments, the heavier mastoid bone and nuchal ridges which are present in the mature male *(vide supra 4.2.1)*. This was corroborated by measuring the mesio-lateral measurement of the condyle as described in *4.5.1*. As some of the skulls had been assigned a gender by the archaeologists which does not fit with the criteria in the table by Mann (1989), the information therein has been used to help resolve the discrepancy.

5.4.2 Age determination (from skull)

If the spheno-occipital synchondrosis was closed it was possible to age the skull as over 25 years. There were also changes that indicated an infant skull or an elderly skull *(vide supra 4.3.1)*. Most of the information was collected from the dentition *(vide infra 5.6.1)*

5.5 Dental Investigations
The investigations carried out are listed below.

5.5.1 Age determination (from dentition)
The statistics were calculated in groups each covering a 10 year range.

Table 4: ***Age groups.***

Group	*AGE RANGE*
1	*0 -9*
2	*10-19*
3	*20-29*
4	*30-39*
5	*>40*

Groups 1 and 2 were assessed by the stage of development using the eruption of the dentition as a guide (Schour and Massler, 1941). Groups three and four were assessed from the general appearance of the dentition and several methods were used. The wear was assessed by Molnar's (1972) scale *(vide infra 5.5.3)* and then the estimation of age at death was calculated by using the comparative table (Figure 11) prepared by Brothwell (1963). He warned that age should not be assessed using the attrition rates from different periods and areas as the diet could change. But, as age was being considered in wide bands of 10 year periods, it was a useful aid if the third band of most severe wear was consulted *(vide supra 4.3.2.9)*. For example, by the age of 20 years +/-3 years there may be 3 or more small pits of dentine exposed in the 1st molar. At 30 years +/- 5 years there may be 3 or more larger pits of dentine. By the late 30s +/- 6 years, secondary dentine may be exposed with no enamel remaining on the occlusal surface and only a thin layer of enamel surrounding the crown (Figure 11).

5.5.2 Teeth present: Ante and post-mortem loss.

The teeth present were charted and missing teeth were noted with all other information in tables as illustrated
The discrepancies in the total tooth number are explained by the postmortem loss and this was noted but not used in the calculations.
The mean tooth loss per dentition was calculated and the significant difference in tooth loss between time periods was examined.

5.5.3 Level of Attrition (Wear)

In each time period the mean wear was measured on mandibular molars only and was assessed for 1st, 2nd and 3rd molars in the same quadrant, if present, using Molnar's scale.

Degrees of molar wear (Molnar, 1972)

1. Unworn.
2. Wear facets.
3. Cusp pattern obliterated. Dentine exposed.
4. 3 or more small dentine patches.
5. 3 or more large dentine patches.
6. Secondary dentine. Tooth surrounded by enamel.
7. Enamel worn away. Extensive secondary dentine.
8. Roots in occlusion.

A cross-tabulation of tooth wear in the different time periods was made to examine significant differences.

Table 5: ***Sample examination chart***

Time period.	**Predynastic**	**Old Kingdom**	**Middle Kingdom**	**New Kingdom**	**Late Period**	**Graeco Roman**
Reference No.	Badarian AF11.5.18	Qau 088	Sudan 270	Qurna 1566	Gizeh E339	Hawara 12888.8.1.62
Teeth Present	*8—4 21/ 1----7* 8---1 / 124--8	*-- / 34 8* 5432/12 7	*6 4/ 3467* 76 / 5	*6—1 / 1—6* 8---1 / 1--- 8	*8---2 /4 6* 87654/ 78	*321 / 23* 3 / 3457
Ante mortem tooth loss	*-- / --* -- / 3	*6--1/1267* 876 / 567	*87 / 8* -- /78	*87 / 78* -- / --	*-- / 78* -- / 6	*64 / 4 5678* 8761 /-
No of teeth	29 present 1 antemortem loss. (a.m.l.)	10 present 16 a.m.l.	9 present 5 a.m.l	28 present 4 a.m.l.	16 present 3 a.m.l..	10 present 11 a.m.l.

5.5.4 Caries

The incidence of caries and the site of attack were noted for each time period. The mean number of teeth affected per dentition was calculated and significant differences between groups were examined.

5.5.5 Examination of the mandibular condyle

Two features of the mandibular condyle were examined and measured.

5.5.5.1 The mesio-lateral measurement

The mesio-lateral measurement of the mandibular condyle was recorded using a Vernier Caliper at the widest points of that dimension. This caliper is manufactured by Tesa, Switzerland. The scale is marked in millimetres and the analogue dial measures to .02 of a millimetre. The measuring blades were placed at the widest points of the mesio-lateral dimension of the condyle and the reading on the caliper was registered on the left side for each specimen if it was intact. If this was fractured the right condyle was measured instead (*vide infra 7.8)*. The reproducibility of this technique was validated as shown in *5.12*. As the gender of many of the skulls in the Duckworth Collection is unknown, those over 21mm and less than 18.5mm in the mesio-lateral dimension were compared with the estimate of gender made by the anatomical appearance of the skull as described in section 5.4. The assessment of gender was corroborated by using the data as explained in 4.5.1. Condylar measurements from child and adolescent skulls were removed from the statistical analysis as they would appear to be female because of their immature size.

5.5.5.2 Form and surface of the condyle

Changes in both the condylar form and surface integrity were measured as an indication of pathological modification or osteoarthritis (Wedel *et al.,* 1978 and Richards & Brown, 1981).

The Form change was graded numerically on the following scale:

- Without changes = 0,
- Slight remodelling =1
- Marked remodelling = 2
- Deforming change = 3.

The Surface change was also graded numerically:

- Without changes = 0
- Uneven surface with an intact, compact layer = 1
- Irregular surface with perforation of compact layer = 2
- Compact layer broken up >3mm² = 3.

The two scales were then added to make *an index of form and surface changes* as an index of osteoarthritic change of the temporomandibular condyle (Wedel *et al.* 1978).

5.6 Periodontal breakdown

As there was no soft tissue to be examined in the osteological collection, this investigation was confined to changes in the alveolar bone. These changes have been recorded as horizontal loss and vertical loss of the supporting bone. There were a few examples of what appeared to be localised aggressive periodontitis (Lang *et al.,* 1999, *vide infra 5.6.3).*

5.6.1 Horizontal bone loss

This was separated from the physiological (apparent loss) which occurs when there is continual eruption of the tooth. If the alveolar crest was uneven, pitted and varied in level in adjacent teeth it was considered to be pathological and was noted. The percentage of skulls affected from each time period was calculated. The condition was noted in the specimens but not measured.

5.6.2 Vertical bone loss

Vertical bone loss which was recognized as an irregular vertical defect or 'ramp' type fault (Hillson, 1996) was also noted and the percentage for each time period was calculated.

5.6.3 Aggressive periodontitis

This can be divided into localised and generalised (Lang *et al., 1999.*) Features of this condition are the circumpubertal onset, localised at the first molar or incisor region and involving very few teeth.

Evidence for localised periodontitis in the material examined is the absence of 1st molars in a small number of the dentitions when the adjacent periodontal condition appears to be healthy. These teeth must have been lost before the eruption of the 2nd molars because the 2nd premolars and the 2nd molars were often in contact and the 2nd molars were frequently tilted. The numbers involved were too small for statistical analysis.

5.7 Continuous eruption and crown height ratio

If the alveolar bone appeared to be intact and at the same level as the surrounding teeth with no evidence of pitting in the crests it was considered that this was an example of continuous eruption. The teeth measured were the lower first molars on the right side. If absent, the lower first molar on the left side was used. No other teeth were selected. The continuous eruption was measured with the Vernier Calipers from the bifurcation of the roots on the alveolar crest to the amelo-cemental junction vertically above. The clinical crown was measured from the alveolar crest at the bifurcation of the roots to the occlusal surface vertically above. A ratio between the two was calculated as an indication of the proportion of continuous eruption to crown height.

5.7.1 Relationship between continuous eruption and crown height and degree of osteoarthritis (F+S)

The relationship between the ratio of continuous eruption to crown height and the degree of osteoarthritis was examined statistically using Spearman's correlation coefficient to see if the ratio increased with the degree of osteoarthritis.

5.7.2 Relationship between continuous eruption and crown height and wear

The relationship between the ratio of continuous eruption to crown height and the measurement for occlusal wear was examined statistically using Spearman's correlation coefficient to see if there was any relationship between the ratio and occlusal wear.

5.7.3 Relationship between the osteoarthritis (F+S) and wear

Osteoarthritis of the temporomandibular joint and the occlusal wear were analysed to see if a direct relationship could be established using Spearman's correlation coefficient.

5.8 Damage attributed to periapical and periodontal disease

This was noted with its relationship to the apex of the roots. Whenever possible, clarification of its aetiology, whether by pulp exposure from caries, wear or as a consequence of periodontal disease, was made. If there is no evidence of these, another possible cause might have been trauma.

5.9 Angle's classification

When both the maxilla and mandible were present, this classification was made from:

- The relationship of the 1st upper and 1st lower molar.
- The overjet, which is the relationship between the upper and lower incisors.

This was possible in 258 of the 524 skulls examined.

5.10 Additional information

Additional information, if available, was noted.

- Anomalies such as supernumerary, malformed or congenitally absent teeth.
- Whether the brain was removed and which pathway was selected.
- In some specimens a mandible has been incorrectly paired with a skull, mismatched by gender and age. These have been noted.

5.11 Hair analysis to determine nutritional status of red haired individuals in Ancient Egypt

This procedure was carried out at and by the Hall Analytical Laboratories Ltd., in Wythenshawe, Manchester using a Fisons MD800 Gas Chromatography-Mass Spectrometry (GC-MS) system.

A very small sample of red hair attached to fragments of scalp, lying loose in a skull storage box from the Old Kingdom period from Nubia, was taken for analysis by a special technique referred to as MSSV-GC-MS was applied. This is Micro-Scale Sealed Vessel pyrolysis – Gas Chromatography – Mass Spectrometry. The MSSV is a small glass vessel, which is used to convert the sample at 300°C into thermal degradation components that can be analysed by GC-MS. Substances can be identified from their pyrolysis products. The presence of additives and contaminants can also be detected which makes it a useful tool in forensic science and particularly appropriate for the analysis of minute samples.

The sample of ancient red hair from the Pearson Nubian Collection was chosen as it could possibly have been discoloured as the result of P.E.M (protein energy malnutrition, Chapter 2), and compared with the data (Thompson, 2001) from another specimen of red hair from an Egyptian mummy from the Bolton Museum and a sample of modern red hair. The skull from the Bolton Museum was dated, by its gold leaf facial decoration and the hairstyle which had been set in curls, to a period from around 664 BC until early AD, The Graeco-Roman Period. The Nubian sample was some 2000 years earlier.

5.11.1 Preparation of hair for analysis

Each hair sample was subjected to the following technique to obtain the Total Ion Chromatograms (TICs) and associated mass spectra by the Micro-Scale Sealed Vessel pyrolysis.

2mg of each sample of hair was sealed in a minute glass reactor with an internal volume of 30-40μl and bent at an angle of 120°. Excess volume in the reactor was filled with pre-cleaned glass beads of 80-120 mesh which were used to reduce the oxygen content of the glass reactor and therefore prevent further oxidation of its contents. Each tube was then sealed by heating the open end of the glass tube in a blowtorch. The specimens were heated in batches in isothermal conditions in a convection oven. Because the mass is so small there is almost instantaneous heat transfer. All samples were heated at

300°C for a period of 72 hours and then fitted to into the inlet of a GC-MS (Garner *et al.,* 1997).

5.11.2 Analysis by Gas Chromatography – Mass Spectrometry

The sample was held in a quantum MSSV port in order to introduce it into the Gas Chromatography section of the GC-MS machine. The glass reactor was opened after the cleaning cycle removed any potential contaminants. The pyrolysis products were released into a stream of helium which is the carrier gas. They were then cryofocused for 30 seconds with liquid nitrogen before being introduced into the gas chromatography column where separation of the individual components was produced.

The initial oven temperature was held at 40°C for 2 minutes. It then was raised gradually to 300°C. The temperature was then held for 5 minutes before cooling. The raw data were recorded from which ion and mass chromatograms were generated. Direct comparisons were then be made with library databases by ionisation at 70eV (Garner *et al.,* 1997).

5.11.3 Hair Analysis by the Scanning and the Analytical Electron Microscope

Two strands of hair were prepared for examination by mounting them and coating them with a fine layer of graphite. The hair was first examined by the scanning electron microscope and was magnified x 30, x 200, x 600, x 900 and x 2,300 to examine the structure of the hair. A hair was also examined by the analytical electron microscope to investigate the inorganic components as an indication of the nutritional health of a specific individual.

5.12 Intra observer variability

In order to access intra observer variability 10 skulls were examined on two separate occasions separated by several months and the results were analyzed for diagnostic agreement.

Using 6 skulls, a further test was devised comparing the readings from two pairs of calipers with three separate measurements; the mesio-lateral measurements of the temporomandibular condyle, the alveolar crest to the amelo-cemental junction and the alveolar crest to the occlusal surface. With one skull the same test was carried out only with the mesio-lateral measurement and the measurement was then repeated with the same pair of calipers on three separate occasions.

5.13 Inter observer variability

Six skulls and two mandibles from the Manchester Museum store were examined by two observers. Each observer assessed age and gender, antemortem tooth loss and wear on the 1st and 2nd molars. As so many teeth were absent because of postmortem damage, an agreement was reached about the teeth which were to be evaluated. Measurements were made with Vernier Calipers of the selected temporomandibular condyle, the height of the clinical crown and the continuous eruption in an appropriate molar. Periodontal breakdown, damage thought to have been caused by abscesses and caries were also noted.

5.14 Statistical Analysis

Data were analysed using the SPSS statistics package (SPSS 10.0. SPSS Inc. 1999). Descriptive statistics involved graphs and tables. Further data analysis was performed using ANOVA, T-tests and Chi-square tests. Kappa statistics and reliability coefficient were used for analysing the inter- and intra-examiner variability.

Chapter 6

Results

6.1 Age

524 specimens in total were examined. 4 skulls were excluded because there were numerous postmortem losses of teeth and 17 mandibles which had no associated skulls were also excluded from the evaluation of age as there were too few landmarks to aid the assessment. 95% (501) of the skulls have been arranged in groups according to age. The range of age in groups 1 and 2 was from 0-19 year, group 3 was 20-29 years, group 4 was 30-40 years and group 5 was considered elderly as described in chapter 5.4. Over 60% of the specimens were from individuals aged between 20 and 40 years at death. Less than a third of the population was perceived to have survived into their 5th decade and it was considered that 15% died before the age of 20. In contrast, in the Graeco-Roman period, the percentage of juvenile deaths had risen to 24%. Age at death for all time periods is listed in Table 6.

There was significant difference (χ^2 test P=0.012) between the groups (age groups 1 and 2 were combined because of small numbers).

6.2 Gender identification

The gender of 91% (476) of the skulls was determined by the anatomical landmarks *(vide supra 5.4.2)*. No gender was assigned to adolescents and children as the characteristically distinctive features differentiating gender do not develop until late puberty. Adolescents and children accounted for 29 of the total. In addition, 19 of the mature skulls were considered to be of indeterminate gender with no significant anatomical indicators. Overall, 265 skulls were concluded to be male and 211 were female.

There was no significant difference (χ^2 test P=0.257) between the groups in gender distribution. (Those of uncertain gender, children and adolescents were excluded). The only group to show any difference was the New Kingdom Group in which 53 were male and 27 were female.

The 75 Graeco Roman skulls examined at the Natural History Museum were more gracile than in the earlier periods and therefore it was difficult to determine gender. As 10 of these had portrait panels in the British Museum collection these were used to corroborate the assessed gender. Seven of these were confirmed to have been correctly assessed. The kappa coefficient was therefore = 0.40 (SE 0.299) and this indicates fair agreement. One skull was excluded as it had not been possible to assess it. These results are listed in Tables 7 and 8.

6.3 Antemortem tooth loss

This was noted for every specimen *(vide supra 5.4.3)*. It was identified by the evidence of healing in the socket and resorption of the alveolar bone. Some sockets contained fractured roots which had occurred because of postmortem damage.

A large number of teeth had been lost postmortem. These were, most frequently, the single rooted anterior and canine and premolar teeth. Many teeth had suffered postmortem damage possibly during excavation or while in museum collections. The mean antemortem loss of teeth varied from a low of 1.76 in the Predynastic group to a high of 4.26 in the New Kingdom. There was a significant difference in tooth loss (P< 0.05) between the New Kingdom and both the Predynastic and Middle Kingdom. There were only two pairs of edentulous maxillae and mandibles.

For antemortem tooth loss see Tables 9 and 10.

6.4 Wear in 1st, 2nd and 3rd mandibular molars

The wear in the 1st, 2nd and 3rd molar was measured as discussed in 5.5.3. Mandibular teeth from the right quadrant were used. If there had been antemortem or postmortem loss, the left quadrant was selected.

Wear has been analysed in two different ways.

1. The mean wear for each time period was calculated and the Anova test carried out. This was calculated for 1st, 2nd and 3rd molars.
2. A cross-tabulation was made between mild wear which was graded using Molnar levels 1-4 and severe wear at levels 5-8 for 1st molars each time period. Chi-square test P<0.001

The mean wear (Molnar, 1972) varied from a low of 3.40 in the Graeco Roman Period to a high of 5.51 in the Middle Kingdom. There were significant differences (P < 0.05) found between the light wear found in the Graeco Roman Period and also in the New Kingdom and the very heavy wear determined in the Middle Kingdom (Molnar, 1972).

For 1st Molar wear see Tables 11 -14.

Similar differences were also found in the 2nd and 3rd mandibular molars.

For 2nd Molar wear see Tables 15 – 17 and for 3rd Molar wear see Tables 18 – 20.

6.5 Condylar measurement

The mesio-lateral measurement was taken in all intact right condyles *(vide supra 5.4.6)*. When these were fractured the left condyle was used.

These ranged in size from a mean of 20.56mm in the New Kingdom to a mean of 19.35mm in the Predynastic and Graeco Roman Period.

See Tables 21 and 22: Condylar measurement.

6.5.1 Gender estimate by condylar measurement

In the Egyptian skulls of known gender the mean for 238 male condyles was 20.94 mm and the mean for the 192 female condyles was 18.92mm. (t-test, $P < 0.001$). Men had a significantly higher condyle measurement. In the 18 of unknown gender the mean was approx 19.6mm. Of these 18, using the condylar measurement for gender estimation, 3 were classified as male as they were greater than 21mm and 5 as female as they were less than 18mm. As the remaining 10 were within the 18mm - 21mm range it was still not possible to classify them.

Within the 245 skulls which were <18.5mm and >21mm there was substantial agreement with gender using anatomical features and condylar measurement. ($P < 0.049$)

The kappa coefficient = 0.648 (SE 0.049; $P<0.049$). This indicates substantial agreement between gender classification and classification using measurement of condyle.

Tables 23, 24 and 25 list the Relationship between gender and condylar measurement.

6.6 Form plus surface measurement as index of osteoarthritic change

This was measured as described in Chapter 4.5.2 using Wedel's scale (1978). The highest mean measurement of change was in the Archaic / Old Kingdom group and the smallest was in the Graeco Roman group which exhibited the least change to the mandibular condyle.

Significant differences in the levels of osteoarthritis ($P<0.05$) were found between the New Kingdom and the Predynastic period and all periods and the Ptolemaic and Roman Period.

Tables 26 and 27 demonstrate osteoarthritic levels during different time periods.

6.7 Caries

The lowest mean for caries was the Predynastic group with 0.23 and the highest mean was in the Graeco Roman group 0.71. There were significant differences in caries ($P < 0.05$) between the Predynastic compared with the New Kingdom and Graeco Roman groups and between the Archaic/Old Kingdom compared with Graeco Roman groups.

The caries incidence is shown in Tables 28 – 30.

6.8 Horizontal periodontal changes

This has to be distinguished from the continuous eruption. (*vide* discussion).

There was no significant difference in horizontal periodontal breakdown between the groups. (χ^2 test $P=0.069$)

Horizontal changes are shown in Table 31.

6.9 Vertical periodontal breakdown

The vertical loss of the alveolar bone which appeared to be the result of periodontal disease was variable in the different groups. In the Predynastic groups it affected 21.1% of the dentitions and in the New Kingdom, 48.3%. There was a significant difference (χ^2 test $P=0.01$) between the groups in vertical periodontal breakdown. See Tables 32 and 33 for Vertical changes.

6.10 Alveolar breakdown

This was interpreted in several ways. Some lesions could be directly related to pulp death resulting from exposure of the pulp caused by wear of the occlusal surface or by caries. Other lesions were difficult to construe and might have been the result of either periapical or periodontal infection and so were considered together. The data from these two pathologies are therefore presented as one cause of alveolar breakdown. Another cause of alveolar damage may be attributed to cysts.

6.10.1 Damage to the alveolar bone attributed to apical or periodontal infection

The data was analysed using the Anova test and no significant differences were noted between any two time periods ($P>0.05$).

See Table 34: Alveolar damage.

6.10.2 Damage to the alveolar bone attributed to cysts

As the number of cysts, which included developmental, radicular and nasopalatine cysts were so few, no statistical test was performed. In 524 skulls there appeared to be bone damage caused by a total of only 18 cysts. More cysts may have been revealed if a radiological assessment had been available *(vide infra 8.14)*. 6 were in skulls from the Archaic/Old Kingdom period and 7 were from the Late Period.
Alveolar damage from cysts is shown in Table 35.

6.11 Dental anomalies, mismatched mandibles, and excerebration through the ethmoid bone

The anomalies noted were supernumerary teeth, absent teeth, malformed teeth or rudimentary teeth, over-erupted teeth, impactions, crowded arches, hypercementosis, taurodontism or trauma e.g. healing wounds. 86 of the skulls exhibited some form of anomaly. Most were apparent in the Late Period.

There was a significant difference (χ^2 test P<0.001) in number of anomalies in the different time periods.
Anomalies are shown in Table 36.

Also noted were mandibles which had been mismatched by archaeologists or incorrectly boxed and for which there was now no possibility of identifying the correct crania. The Predynastic and Middle Kingdom were correctly assigned but 15 mandibles from other periods had been mismatched. Mismatched mandibles are shown in Table 37.

No example of excerebration was noted earlier than in the New Kingdom in this investigation. In this survey it was observed in 55% of the skulls in the New Kingdom, 16.7% of the skulls in the Late Period and 13.5% of the Graeco Roman skulls. There was a significant difference among the three groups (χ^2 test P<0.001). Table 38 illustrates distribution of excerebration.

6.12 The relationship between age and wear

As would be expected, there was a direct relationship with increasing age and an increase in wear of the occlusal surface of the molars. Spearman's correlation coefficient = 0.745 (P < 0.001)
The relationship between wear and age is listed in Table 39.

6.13 Relationship between osteoarthritis and age

There was an increase in the measured osteoarthritic changes of the temporomandibular joint with age when analysed by Spearman's correlation coefficient = 0.458 (P < 0.001)
The relationship between osteoarthritis and age is illustrated in Table 40.

6.14 Relationship between wear in the 1st molar and osteoarthritis

A significant relationship was found between the wear of the 1st molar and the degree of osteoarthritis.

Correlation is significant at the .01 level (2-tailed) when analysed by Spearman's correlation coefficient =0.448.
The relationship between wear in the 1st molar and osteoarthritis is shown in Tables 41 and 42.

6.15 Measurement of continuous eruption

As was explained in 5.7 the measurement was made from the alveolar crest at the bifurcation of the roots of the 1st mandibular molar vertically to the amelo-cemental junction as a measure of the continuous eruption of the tooth when there was no suspicion of periodontal disease. The results were subjected to the Anova test and no significant differences were found between any 2 groups (P>0.05).

Table 43 lists the measurement of continuous eruption.

6.16 Relationship of the ratio of continuous eruption and crown height to wear of 1st molar

The ratio of the continuous eruption to the crown height was measured to determine whether wear in the 1st molar influenced continuous eruption. The results were analysed using Spearman's correlation coefficient = 0.444 (P<0.001). This relationship is highly significant.
Continuous eruption and crown height to wear of 1st molar is listed in Table 44.

6.17 Relationship of ratio of continuous eruption and crown height to measurement of Form + Surface (F+S) of the condyle (Measurement of osteoarthritis)

The measurement of the ratio of the continuous eruption to the crown height was compared with the level of osteoarthritis to determine whether there was a direct relationship between them. The results were analysed using Spearman's correlation coefficient = 0.318 (P<0.001). A significant increase in ratio of crown height and continuous eruption was found with increasing levels of osteoarthritis of the temporomandibular joint.

Continuous eruption and crown height measurement to osteoarthritis is shown in Table 45.

6.18 Angle's Dental relationship

It was possible to determine the dental relationship in 258 of the skulls examined either by the molar relationship or by the incisor relationship (vide supra 5.10). 47.3% had a Class I relationship, 16.7% had a Class II relationship and 36% had a Class III relationship.

See Table 46 for the comparison of Angle's relationships.

6.19 Hair analysis by Gas Chromatography-Mass Spectrometry

A small sample of red hair from a Nubian male from the Old Kingdom, still attached to the skull by flakes of scalp, was taken, with permission, from the Duckworth Collection in Cambridge for analysis by the Micro-Scale Sealed Vessel Pyrolysis - Gas Chromatography-Mass Spectrometry (MSSV-GC-MS) technique *(vide supra 5.12)*. This analysis was carried out to see if the red hair might be an indication of protein energy malnutrition (P.E.M) as this could have been one of the factors in the aetiology of the periodontal disease which was in evidence in many of the dentitions examined in this study.

The GC-MS data thus generated are represented in several ways (Table 47). The horizontal axis of the chromatogram measures the time period and the vertical axis, the relative intensity of the peak. Above the vertical axis are the sample number, type of analysis, the temperature and the time period for that analysis. The Total Ion Chromatogram (TIC) of the Nubian hair showed many peaks. Each peak denotes a different chemical component. A mass chromatogram shows a sequence of peaks representing a particular mass to charge (m/z) ratio.

The more volatile components were released from the chromatographic column more quickly than the less volatile components which tend to have a heavier molecular weight. The peaks formed in 0-15minutes were small, thermally formed fragments. In 20-25 minutes were peaks of C_{14}-C_{16} and, after 30+ minutes, peaks of C_{30} appeared. Each peak has its corresponding mass spectrum.

Two further chromatograms were compared. These were from a sample of red hair from an Egyptian Mummy from Bolton Museum and a sample of modern, naturally red hair (Table 48).

Specific ions were selected for analysis. Chromatograms of long chain fatty acid nitriles, which are formed from the lipids in hair, were compared for each of the three samples. These are diagnostic for the lipid profile of a particular hair sample. The mass chromatograms for m/z = 97 for these three samples were evaluated. When the 17-25 minutes region was compared for the three samples of hair there were similarities found between the two ancient Egyptian specimens but they differed from the modern hair (Table 49). This indicated that the ancient hair had experienced degradation. The relative peak height ratios for G_{15} and G_{16} fatty acids between 20-23 minutes are diagnostic and these may be used to distinguish between ancient and modern samples.

The trace from the Nubian for G_{16} at 23 minutes was then compared with a library specimen which was a good match (Table 50). Other peaks can be identified by comparing the corresponding mass spectra with library data which contains 30,000 samples in its data base. The best matches can be listed in a matter of seconds.

After 30 minutes the peaks are thought to be the long chain esters (C_{20}) of long chain fatty acids (C_{16}) which are characteristic of beeswax and are shown in Table 51 (vide infra in Discussion).

Table 52 shows the mass chromatogram for m/z = 217 diagnostic of steroids. In the top figure the peak at 32.04 is matched with the library traces below. The lipid profile shown in Table 53 and the steroid profile above show that the red-haired Nubian was nutritionally in good condition with no indication of protein energy malnutrition.

The hair was then scanned by electron microscope at the University of Manchester.

At a magnification of x 30, it showed 3 hairs in a flake of scalp. x 200 the specimen was covered with a crystalline substance which was identified as hydrous calcium sulphate. Where the shaft of hair emerged from the scalp there were no scales apparent. The scales became perceptible at magnifications of x 600 and x 900. Some small round grains were detected and when they were magnified x 2300 it was found to be a pollen grain. When the hair was examined by the analytic electron microscope the inorganic components showed no evidence of malnutrition. The hairs are shown in Figures 41a (x 30), 41b (x 200), 41c (x 900) and 41d (x 2300).

6.20 Reliability

6.20.1 Intra-observer reliability

Within the parameters measured for intra-observer variability, using the Kappa Coefficient or reliability coefficient, it was found that 'gender' showed substantial agreement at 0.737 but all others rated almost perfect as they were greater than 0.81.

Intra-observer reliability is listed in Table 54.

A further test was carried out using different calipers on 2 separate occasions to compare the results of measuring the condylar size, continuous eruption and clinical crown

height of 6 specimens. Reliability coefficients show perfect agreement as they were all higher than 0.99.
Intra-observer reliability with calipers is shown in Table 55.

6.20.2 Inter-observer reliability

The researcher and supervisor both examined 6 specimens. Using the Kappa Coefficient for agreement, it was found that there was perfect agreement for gender, age and antemortem loss, osteoarthritis (F+S) and periodontal condition. There was almost perfect agreement with measurements for wear in the 1st molar, condylar width, continuous eruption, clinical crown height and caries. There was substantial agreement for wear in the 2nd molar. In the interpretation of bone destruction which indicated periapical and periodontal abscesses there was fair agreement.

Inter-observer reliability is listed in Table 56.

The significance of these results is discussed in the next chapter.

Chapter 7

Discussion

7.1 History of the collections.

The provenance, the acquisition and the estimated time period covered by the collections from The Duckworth Collection in Cambridge and the Petrie Collection in the British Museum (Natural History) is described below.

7.1.1 Predynastic collections (5000-3100 BC)

The following description of Sequence Dating explains how the age of the collections is confirmed.

Flinders Petrie, when working on the pottery excavated from Naqada, had compiled a method of Sequence Dating (S.D.) which is still the accepted method of ascertaining age in the predynastic period. He assigned a number to what, he considered, was an unquestionable development in a change of design of handles in pots from wavy-handled pots to ornamental appendages. Petrie used the pots found in the Royal Tombs of Abydos from the 1st Dynasty as his base line. He worked back to the earlier predynastic period of Naqada 1 to which he assigned the S.D. 30 so that any future finds from earlier periods could then be placed in that sequence. He considered the pottery together with the tomb objects surrounding them to help him establish their chronological dating in 900 of over 4000 tombs which he had excavated between 1899 and 1903 (Drower, 1985). Their strata in the tomb further refined the dating. The final number of 77 was given to the 1st Dynasty (Gardiner, 1961).

The site at El-Badari was excavated by Brunton and Caton-Thompson in the 1920s and, although it is impossible to date these predynastic periods with complete accuracy, the Badarian era is considered to be the earliest predynastic period in the group. The Badarian remains are earlier than 30 and therefore the Sequence Date that has been assigned to it is 21-29. Naqada has been divided into two periods: Naqada 1which has a S.D. of 30-39 and Naqada 2, a S.D. of 40-62. Hierakonpolis has a S.D. of 35 (Gardiner, 1961).

An example of the difficulties encountered in finding adequate numbers of research specimens for examination is illustrated by the Naqada collection. This was listed as containing 500 skulls, but only 91 were reported to be 'E', indicating that the skull is entire. Within this group, several had the whole dentition missing because of postmortem loss. There were several with absent maxillae, one with a mismatched mandible and many were disarticulated. In consequence, of 500 skulls from this period, only 40 were considered satisfactory for this study.

The sample size was chosen because, in a previous study of severe tooth wear (Miller, 1997), there was a prevalence of 57% of severe tooth wear at level 5-8 on the Molnar 1971 scale (*vide supra 5.5.3*). The sample size calculation (Elashoff, 2000) showed that a one-group chi-square test with an α = 0.05 two sided significance level would have 88% power to detect the difference between the Null hypothesis proportion of 60% prevalence of severe wear and the Alternative proportion of 40% severe wear when the sample size was 60. It was decided that a minimum of 60 skulls from each time period would be examined depending on their availability.

Specimens from El Badari , Naqada and Nubia were examined.

7.1.2Archaic and Old Kingdom (3100 - 2613 BC)

The Abydos collection is probably from the Archaic Period, which encompasses the 1st and 2nd Dynasties, from 3100 –2686 B.C. Amelineau, who admitted smashing any incomplete artefact that he considered of little monetary value, excavated Abydos from 1895 (Drower, 1985). Flinders Petrie then took over the concession, continued with the excavations and built up a list of the early kings based on the differences in style of the grave goods and on early Hieroglyphs on plaques and seals which had been ignored by Amelineau (Drower, 1985).

There were Archaic Period burials at Abydos in Middle Egypt, which, with Memphis, was a centre of power during this period. There are also cemeteries in Nubia, in Upper Egypt, and the Delta and Saqqara and Tarkhan, near Gizeh in Lower Egypt, which spanned the period from the Archaic Period to the Old Kingdom (Baines & Malek, 1958).

During this period mastaba tombs, built to protect the body of the deceased from being disturbed, were used only for the pharaohs and nobles. The lower classes were buried in the hot sand of the desert. The mastaba superstructure was made of mud brick and was rectangular in shape. The burial was in the substructure which contained several rooms and store rooms for the funerary goods. In the 3rd Dynasty of the Archaic Period the pharaoh, Djoser, built the first pyramid, the Step Pyramid, for his burial. The Step Pyramid, which began as a square rather than rectangular mastaba, was made of stone instead of mud brick and is the earliest stone building recorded. Each of its steps is in the form of a smaller mastaba, and together they form a pyramidal shape. The architect of the pyramid was Imhotep, also recognised as a writer and a healer who was later identified with the Greek god of healing, Asklepios (Gardiner, 1961). By the 4th Dynasty the smooth sided true pyramid had developed (Watson, 1987).

Grafton Elliot Smith, who had been Professor of Anatomy at Manchester University, examined the human remains from the cemeteries of Nubia in 1907 as part of the First Archaeological Survey of Nubia. He had an acrimonious relationship with Petrie over the condition of some of the mummies that, Petrie claimed, had been '*dissevered*' or disarticulated and maintained that this was a custom of the '*higher classes'* at the beginning of the Pyramid Period. Smith disagreed, saying that he had examined thousands of mummies and had never seen any evidence of this. It is now thought that these bodies were dismembered accidentally because they had been disturbed in the past and had been reburied or re-wrapped in antiquity.

From a photocopy of Elliot Smith and Wood Green's original handwritten list supplied for the Nubian collection at Cambridge, it can be seen that the graves had been used for multiple burials from predynastic times to the Christian era, a period of some 3,500 years. There was, therefore, difficulty in identifying the relevant skulls to be included in this first survey of predynastic dentitions.

Skulls from Abydos, Qau, Nubia (Pearson Collection), Tarkhan, Dishasha and Aswan were examined for the research reported in this thesis.

7.1.3 Middle Kingdom (1991 –1786 BC)

Although there was a history of famines during this period, this era was also characterised by the reorganization of the country by Amenemhet 1, the 1st pharaoh of the 12th Dynasty. Amenemhet 1 said that '*he had banished poverty and that during the famine, no one was hungry.'* (Gardiner 1961). This quotation is reminiscent of the role of Joseph from the Old Testament. It also fits in within a hundred years of Rohl's (1995) speculative theory about the timing of Joseph's presence in Egypt. This, however, is not the generally accepted date for Joseph who may have been a vizier during the 2nd Intermediate Period when the Hyksos were rulers. Kerma, which lies to the south of the Third Cataract, indicates the extent of Egypt's control during the Middle Kingdom. The border of the Middle Kingdom territory was at the Second Cataract and it is possible that Kerma was a southern trading post. At the burial site, the dead were not mummified but laid on their sides. Their wives and servants had been killed and buried to accompany them in the afterlife (Gardiner, 1961). During the Old Kingdom, Thebes was only a village but, during the Middle Kingdom, it became the second capital. El Tarif was a site opposite Thebes on the west bank. Specimens from Kerma and Qurna el Tarif were examined for this thesis.

7.1.4 New Kingdom (1567-1085 BC)

There were specimens from excavations at the Ramesseum, Memphis, Saqqara, Qurna and Gurob in the New Kingdom Collection. From this period, skulls from sites in Qurna, from Memphis and from Saqqara were selected. These were areas of prosperity at that time. Thebes was the religious capital in the 18th Dynasty and Qurna is a site opposite on the west bank of the Nile. Memphis was the administrative capital from the late 18th dynasty with a large foreign population. Saqqara is a site of historical significance near Memphis where government officials, priests, craftsmen and palace employees were buried (David, 1999).

7.1.5 Late Period (664 –332 BC)

Gizeh, the source of the Late Period collection, was excavated by Reisner who was the professor of Egyptology at Harvard. He set out principles for systematic excavations (David, 1999). This means that each stratum is excavated and its contents are carefully noted in inverse order. After the Old Kingdom, Gizeh's significance had diminished and only became important again from the New Kingdom. This included a period of Nubian rule but the country was less independent economically and relied on trade with the Near East. By the 27th Dynasty there was a foreign ruler, Cambyses, who was the first of the Persian pharaohs.

7.1.6 Graeco-Roman Period (332 BC – 400 AD)

Alexander the Great conquered Egypt in 332 BC and after his death the rule passed to a general, Ptolemy. There was then an influx of Greeks to Egypt and Greek became the 'lingua franca' of government but Hieroglyphs continued to be used for religious writings (David, 1999). By 30 BC Egypt became part of the Roman Empire. The few available skulls of that period from sites in Nubia, Memphis and Gurob in the Duckworth collection were augmented by skulls from Hawara in the Natural History Museum collection.

7.1.7 Intermediate Periods

No examination of skulls from the 1st, 2nd and 3rd Intermediate Periods was made as these were unsettled periods where the influx of invaders from other countries did not necessarily reflect the dental health of the indigenous population.

7.2 Skull anomalies

A general appraisal was made of the skull and any found anomaly was noted.

7.2.1 Metopic sutures

At birth the frontal bone consists of two halves separated by the metopic suture which extends from the glabella to the anterior fontanelle. This begins to fuse at about 2 years after birth and has usually closed completely by about eight years. There were few examples of metopic sutures in the 524 the skulls examined. Those noted were 1 persistent suture in a skull from Gurob (Figure 12) during the New Kingdom, 1 from Gizeh from the Late Period and 4 examples in the Graeco-Roman skulls from Hawara. In an additional sample taken of British Neolithic skulls, 3 patent metopic sutures were discovered within the 14 examined.

7.2.2 Wormian bones

(Figure 13). These are found in the lamdoidal suture and are known as ossa suturarum. They are often symmetrical but have no morphological significance (Johnston *et al.,* 1958). These were noted in only a very few crania.

7.3 Skull pathology

There is some evidence of metabolic disease, benign tumours, possible cases of malignant neoplastic disease and healing wounds. Osteoarthritis will be discussed later.

7.3.1 'Pseudopathology' or pitfalls in the examination of crania

Wells (1967) has written about the *'pseudopathology'* which might cause incorrect conclusions being made about skeletal material. The danger is that diagnoses of pathologies might be made where there are none.
There are many causes of mistaken diagnoses:

7.3.1.1 Physical

The buried skull can be deformed by the weight of the soil when it is exposed to conditions which may cause the demineralisation of bone (Brothwell 1981). Deformation does not occur when the body has been mummified because it is has been desiccated and then placed in a coffin and sarcophagus and is thus protected from the weight of soil from direct burial. However, although teeth are the most stable part of skeleton as they are the hardest and most chemically stable tissue in the body (Roberts & Manchester [2] 1995) the shrinkage of dentine may cause the fracture of the dental enamel which then may be mistakenly thought to have occurred because of the undermining of enamel by caries. Fenestration in the alveolus may appear to be pathological but might have been caused by postmortem wear. There may be the postmortem loss of teeth because of degeneration of the periodontal membrane and the careless handling of the material.

7.3.1.2 Chemical

Chemical erosion of the bone can cause pitting which may look like cribra. Osteolysis may resemble osteitis or osteomyelitis. Physico-chemical deposition of precipitated carbonates is common and may be mistaken for calculus (Wells, 1967).

7.3.1.3 Animal damage

Beetles can cause holes in bone and in skulls which may be mistaken for myeloma. Rodent damage has been found around orbits (Wells, 1967). No examples of these conditions were found in the Ancient Egyptians but there was one example in the British Neolithic skulls of, what appeared to be, defects not dissimilar to that caused by myeloma (Figure 19).

7.3.1.4 Excavation hazards

Post mortem fractures of bones are a common occurrence. There are also fifteen examples in this survey of the incorrect mandible being assigned to the skull. Several of these have been mismatched both by age and by gender (Figure 14).

7.3.2 Valid pathology of the skull

7.3.2.1 Cribra orbitalia

This is a cribrous area in the roof of the orbit. A similar condition can often be seen in the parietal bones. It is thought to be the response to over activity of the bone marrow resulting from its stimulation by anaemia (Figure 15). There are three types of anaemia which might be implicated. These are iron deficiency anaemia, sickle-cell anaemia and thalassemia. Roberts and Manchester (1995) suggested that the cause of iron deficiency anaemia could have resulted from the transition from a hunter-gathering economy to agricultural economy since the phytates in some cereal crops inhibit iron absorption.
Satinoff (1972) observed that skull radiographs showed the 'hair on end' appearance in both sickle-cell anaemia and thalassaemia but neither of these pathologies has been confirmed. In both of these conditions the medullary layer of the orbital plate of the frontal bone expands as a response to the extra demands on the marrow which has become hyperplastic. The surrounding tables of cortical bone thin and erode. There is new bone formation, the cribra (Gilles, 1989). Contemporary populations of East Africans have a high incidence of sickle-cell anaemia which could have been the cause of the cribra seen in skulls from Nubia, in southern Egypt, during the Old Kingdom. As there is a high incidence of thalassaemia in the Middle East in the present day, this might explain the incidence during the Late Period in Gizeh, which is nearer the Mediterranean. Alternatively, and in both regions, it could be the result of iron deficiency anaemia. It may be possible to analyse any hair left attached to the occasional skull to assay it for

iron by mass spectrometry. As it is not possible to see cribra in the intact mummy that plentiful source of hair would not be useful.

One other medical condition which may be implicated in producing cribra is scurvy (Ortner, 1984, Roberts, 1987). Ortner examined a skull from Metlatavik in Alaska with cribra. In this case, because of the geographical location, there is no possibility of this condition being caused by either thalassaemia or sickle cell anaemia. He reported that the anterior teeth were missing and that there was '*evidence of reactive bone in the alveolar cavity of the incisors which suggests antemortem loss with some attempt to remodel the cavity.*' In scurvy there is a deficiency of vitamin C, which is necessary for the formation of collagen. As a result, the periodontal membrane is defective and so antemortem loss of teeth, especially single rooted teeth, is highly likely. As Ortner (1984) suggested that the age of the child was eight years old, it has to be considered that the root formation of these teeth would not have been completed. Therefore, what appeared to be healing in the socket might well have been the consequence of the shorter length of the socket found when there is incomplete root formation and it was thus an indication of postmortem not antemortem loss.

Researchers, interviewed for the 'Discovery' TV programme on 'Mummies' on 13th May 2001, when examining the cribra found in skulls in the Turin collection, mixed ground dry bone from some specimens for the immunological detection of malaria and found that 40% were positive. Sandison (1967) was of the opinion that malaria may explain the finding in ancient records which relate that the Ancient Egyptians suffered an annual fever. There is an inscription in the temple at Denderah in Middle Egypt to that effect. Herodotus wrote that '*the country is infested by swarms of gnats, and the people have invented various methods of dealing with them; south of the marshes they sleep at night on raised structures, which is a great benefit to them because the gnats are prevented by the wind from flying high; in the marsh-country itself they do not have these towers, but everyone, instead, provides himself with a net, which during the day he uses for fishing, and at night fixes up around his bed and creeps in under it before he goes to sleep. For anyone to sleep wrapped in a cloak or in linen would be useless, for the gnats would bite through them; but they do not even attempt to get through the net'.*

7.3.2.2 Biparietal depressions

This is a condition of unknown aetiology but several factors are cited such as old age and a possible relationship between osteoporosis and parietal thinning. It was first described by Elliot Smith in 1907 as '*strange, large symmetrical depressions of the parietal bones'*. He said that it never occurs in parts of skull covered by muscle and he thought that it was caused by pressure (Lodge, 1967). Bilateral thinning was usually found at the posterior part of the parietal bones. The inner table of bone is normal but there is a loss of bone in the outer table and diploic marrow cavity. Satinoff (1972) found no link between age and gender in the skulls examined. He described the appearance as '*carpetbag*' or '*Gladstone bag*' in form. His radiographs show a thinning of the outer table of the parietal bone. Several examples of this condition have been found in the Gizeh skulls of the Late Period (Figure 16).

7.3.2.3 Osteomata

These were found on the frontal bone of several of the skulls (Figure 18) Osteomata are usually found in the 40-50 years age group and they are well-defined radio-opaque protuberances of dense, mainly lamellar bone (Muir, 1992). The lesions are benign, smooth and dense and symptomless. Brothwell (1967[2]) considered them to be discontinuous variants of anthropological interest and non pathological.

7.3.2.4 Neurofibromas

This lesion may be isolated or occur as multiple tumours, neurofibromatosis. It is a developmental disorder of the nerve sheaths and there is often a family history. If the cutaneous nerves are involved the skin may have melanin pigmented tumours known as '*café au lait*' spots. A high incidence of malignant change in occurs in neurofibromatosis. It is very rare in solitary neurofibromas. There are oral manifestations in almost 7% of the cases and the occasional interosseous sites are more common in the mandible at the inferior dental (Figure 17a) and mental foramina (Figure 17b) than in the maxilla (Soames & Southam, 1998). Three possible examples of this have been found in the Kerma Sudan group from the Middle Kingdom. Although it is unknown if there was any genetic link, DNA studies of these three skulls could be undertaken to investigate whether the skulls came from related individuals.

7.3.2.5 Multiple myeloma

This is a neoplastic tumour occurring mainly in the 50-60 years age group. The multifocal lesions originate diffusely in the plasma cells of the bone marrow. The bone resorption is caused by tumour secretion of osteoclast activating factors (Muir, 1992). Such lesions must be differentiated from the metastases from other tumours (Roberts & Manchester, 1995). One suspected case in the survey of the additional British Neolithic material has been found in a skull which was assessed as elderly. Wells (1967) said that holes, which might be mistaken for myeloma, could be caused by postmortem beetle damage. Molleson (2001) discussed a Nubian skull in the NHM with holes which may be myeloma or a postmortem attack by ants.

7.3.2.6 Cleft Palate

No examples of this condition were found in the 524 skulls examined. As the contemporary incidence is 1 per 600 births (Ferguson, 1978) this is not unexpected. Cleft lip and palate are known to have a genetic cause but, with cleft palate alone there is less of a family history and it may be environmental. However, the only report of this condition found in Ancient Egypt is of a defect in an adult of the 25th Dynasty (Roberts & Manchester 1995). The rarity of this condition in palaeopathology may be because the infant died shortly after birth and the mainly cartilaginous skull would have decomposed. The survival of an infant would be dependent on adequate nutrition and feeding and would have taken time and ingenuity. Ferguson (1978) also suggested that, historically, infanticide may have been practised on the abnormal. Gilkey (1978) reported 3 cases which were all pre-Columbian.

7.3.3 Wounds

Several skulls were found with wounds to the frontal bone which showed signs of bone regeneration and were therefore not the immediate cause of death. These may have been injuries from battles or work associated trauma from the massive building projects which abounded in Egypt in ancient times. Please see Figures 20a and 20b. Figure 20c shows a healed wound to the underside of the mandible.

No skulls examined for this study showed evidence of trephination. Lisowski (1967) discussed six examples from Ancient Egypt. He considered that this was remarkably few as more Egyptian skulls have been examined than of any other population. Two of the six were from the New Kingdom c. 1200 BC, one from c. 600 BC, one from the Ptolemaic period c. 323 –30 BC, one from the Roman period c. 50 – 200 AD and one from the Byzantine period c. 395 – 638 AD. Lisowski thought that the Ptolemaic specimen had been influenced by 'Greek' surgery from that period and the last two, by 'Roman' surgery. His photograph of the skull from the Ptolemaic period displayed only one side but the skull is reported to have been trephined on both sides and also shows evidence of parietal thinning.

7.4 Age at death

Because of its abundance of food, fresh water and the materials to construct shelter Harer (1993) compared the Nile Valley to the Garden of Eden. Despite this, the average life span was short. Waldron (2001) wrote that the peak death periods were perinatal, between birth and puberty and in old age. The early death rate of females was because of the complications of pregnancy and labour. Males were perceived to have an easier life than females but were more likely to suffer death from trauma. Other causes of early death might be conjectured to result from dehydration caused by drought, starvation, tetanus from soil contamination of wounds, the sequela of infestation by parasitical worms and possibly epidemics of gastroenteritis, measles and malaria.

The average age at death in the different time periods varied. Age has been calculated from examination of the skull according to the eruption times of the teeth, the closing of the spheno-occipital synchondrosis and an assessment of the degree of occlusal wear (*vide supra 5.4.1*). Another system of age assessment utilised Brothwell's table (1994) of comparison of differing severity of wear patterns exhibited by the molars at different ages. Cottone & Standish (1981) stated that age estimate in the first decade is +/- 1year, in the second decade it is +/- 2 years. Therefore the older the age estimate the greater the range of error.

In this study, the highest percentage of child and adolescent deaths was during the Graeco Roman Period. 17 of the 75 skulls examined were considered to be under the age of 20 years at the time of death and 4 of these were estimated to be less than 10 years. These juvenile deaths represent such a high proportion of the population that they may have been prone to the illnesses discussed earlier. Because an individual cemetery may have been in use for several hundred years these skulls may not have been a representative sample for that time period and population and this may have affected the estimated average age at death (Roberts and Manchester, 1995). For example, a part of an excavated cemetery may have been used only for the young.

7.4.1 Malnutrition

The Malthusian Theory states that the population growth will always tend to outrun the food supply (Encyclopaedia Britannica, 2002). As the population changed from hunter-gathering to become settled agriculturists with increased dependence on cereal crops, it was at the mercy of the annual inundation. The cause of malnutrition in ancient Egypt resulted from shortages of food because of drought, flooding or infestation of locusts. The reduced intake of protein from hunting would also have contributed to malnutrition. Scurvy, as has already been discussed (7.3.2.1), is one of the diseases thought to be responsible for cribra orbitalia (Ortner, 1985). Other causes of porotic hyperostosis, which was found in the Nubian population, is iron deficiency anaemia which is most common in children and young females as cereal is poor in iron and calcium (Armelagos & Mills, 1993). Frequent childbirth caused problems of nutrition for the older child when there was a younger one still dependent on breast feeding. The elder child was '*displaced from the breast*'. This is one of the causes of kwashiorkor (*vide supra* 2.7). Other signs of malnutrition or chronic illness have been found in the long bones as lines of density, Harris lines, which indicate arrested development. Hypoplasia in the dentition may also denote malnutrition as well as other causes of childhood stress such as infections (Figure 21).

7.4.2 Epidemics

Mortality and morbidity caused by micro-organisms far exceeds that due to warfare and famine (Roberts and Manchester, 1995). There is an association between community size and infectious disease, e.g. measles, has been studied in modern isolated societies. In small communities a virus will infect, kill or immunize. The virus dies out and will only reoccur if reintroduced from outside. Acute infections do not leave skeletal evidence. Specific diseases will affect the sections of the population differently depending on their age, sex or genetic profile and this is also influenced by diet and living conditions. Herd immunity may develop, induced by repeated infection building up antibodies within the community, thus conferring resistance. Infants and young children are susceptible to respiratory disease and gastroenteritis. There may have been outbreaks of malaria as its antigen has been found in mummies from all periods (Nunn, 1996).

7.4.3 Disease

Thalassaemia was prevalent around the Mediterranean and, further south, the Nubians and Sudanese may have suffered from sickle cell anaemia which would have shortened the lifespan. These conditions, conversely, confer a resistance to malaria because the infection cannot develop because of the rapid turnover of red blood cells (Roberts & Manchester, 1995). This may have been the reason why 8 of the 60 skulls from Gizeh, in the Late Period, had cribra. Tuberculosis was another disease that may have occurred in ancient Egypt. The mycobacterium is believed to have evolved from a saprophytic soil bacteria which affected cattle and then changed into the human form of the organism (Waldron, 2001). Agriculture started during the Neolithic period while some groups of the population began the domestication of animals. Milking of cattle is recorded from the fifth millennium BC. There is evidence from the Old Kingdom, in the 6th Dynasty tomb of an official called Nekhebu, depicting the cultivation of the fields and the separate pasturing of flocks and cattle (Eyre, 1999). Pott's disease has been recorded by Ruffer and there are cases recorded from the Predynastic period onwards (Nunn, 1996). Parasitic diseases were common and many would have died from schistosomiasis, dracunculiasis, filiariasis, and strongyloidiasis (Tapp, 1979 and Tapp & Wildsmith, 1992).

7.4.4 Trauma

Some skulls showed evidence of trauma, probably sustained in battle. Several of these showed considerable healing and so were probably not the immediate cause of death (Figures 20a, 20b and 20c). There are also examples of the healing of long bones after fractures and splinting was discussed in the *Edwin Smith Papyrus* (Breasted, 1930). Compound fractures which did not heal may have caused death by septicaemia or as the sequelae of osteomyelitis. A common cause of trauma and death seems likely to have been a crocodile attack. The *Ebers Papyrus* states '*What has to be done for the bite of a crocodile. If you examine the bite of a crocodile and you find it with his flesh piled up and its two sides being separated, then you should bandage it with fresh meat on the first day, likewise all wounds of a man'* Those drowned during an attack were called the 'blessed drowned ones' (Nunn, 1996).

7.4.5 Poisoning

Deaths may also have been caused by toxins from cobra, vipers and scorpions which were part of the Nile fauna. Some mummies have been found to contain a high level of lead. One of its sources may have been the galena, a powdered mineral, which was mixed with water and was used from the Predynastic Badarian period to enhance the eye as well as to discourage the numerous flies which transmitted eye disease (Dayagi-Mendels 1989). Also a lead drinking straw has been found from the New Kingdom period in Amarna. There are no natural deposits of lead in Egypt (Harer, 1993).

7.4.6 Malignancies

Malignancies do not appear to have been common, possibly because of the brevity of life. The changes in the tissue caused by mummification make evidence of soft tissue tumours difficult to detect. In skeletal tissue, suspected tumours must be differentiated from pseudo pathology such as beetle damage. Primary tumours which have been considered to be multiple myelomatosis may be secondary tumours from metastatic spread (Strouhal, 1976). He also noted that the prevalence of skull tumours found may have reflected the availability of skulls and preference for them by previous anthropologists. Waldron (2001) considered the estimate of those dying from malignancies to be an underestimate.

7.5 Gender

7.5.1 Gender from the skull

Sexually dimorphic traits may vary in different population groups. The skull is less reliable than the pelvic girdle for determination of gender. The percentage of one gender at a specific site might not reflect the true proportion of the genders in the population as only a part of a burial site may have been excavated. If all the skulls found were male, it may have been reserved specifically as a workman's site. According to Mann (1990), there were considerable discrepancies in the original excavators' sexing of Badarian skulls from the Predynastic Period and in Qau from the Old Kingdom. All those which had been previously assigned a gender were therefore reassessed. After this reassessment of the mature skulls, 19 of the 524 were of uncertain gender. However, as 10 of the 19 were in the first group which were examined, it is now thought that the problem was an initial diffidence in this researcher to contradict the original excavators' sexing of the collection.

As was described in 4.2.1, no attempt was made to ascribe gender to adolescents and children. Those represented 29 of the total.

The 58 Graeco Roman skulls from Hawara were very difficult to classify as they were mostly gracile and did not have the defined anatomical features found in most of the other skulls. However, 10 of these had had panel portraits included within their bandaging when they had been excavated and the portraits are now part of the British Museum Egyptology Collection. The paintings, from Hawara, are exceptionally fine examples of the Fayum portraits. During that period in the 1st Century AD, instead of using stylised coffin images or cartonnage facemasks which had been used during the earlier eras, a portrait on wood or linen was painted with encaustic or tempera during the subject's lifetime (Doxiadis 1995). After death, it was cut down and included in the elaborate bandaging of the mummy used in that period. Figure 22 is the portrait of one of the skulls with which there was ambivalence in the classification of gender.

In the Roman period, there was a marked deterioration in the technique of mummification and so that the resulting preservation of the bodies was poor but the bandaging and ornamentation, which had become more elaborate, appeared to be of prime importance. Susan Walker, the Deputy Keeper in the Department of Greek and Roman Antiquities of the British Museum, provided the information about the gender of the portraits. The assessment made from the examination of the skulls was in agreement with the genders of 7 of the 10.

7.5.2 Tooth gender

Although there is a difference in eruption times between males and females as was discussed in Chapter 4 (with the female dentition erupting consistently earlier) it was not possible to use this information to help in the determination of gender. Hillson (1996) reported sexual dimorphism in crown size but that there was only a very small difference of 0.4 - 0.5 mm and therefore likely to be a source of inter- and intra- observer error. Crown size was not used in this survey.

7.6 Antemortem tooth loss

Teeth may be absent, having been lost ante- or postmortem, having failed to erupt, or by being congenitally absent. It is not always possible, without radiological verification, to be certain whether teeth are unerupted or congenitally absent. As a general rule, an alveolus containing an unerupted tooth tends to appear expanded.

There are many causes of antemortem tooth loss in antiquity.

Cottone and Standish (1981) observed that although the '*dentition was subject to breakdown throughout life; it outlasted all other body tissues after death*'. Brothwell (1963) thought that the aetiology of tooth loss had changed in different locations and time periods. The causes were wear, pulp exposure and periodontal involvement. Hartnady and Rose (1991) examined 104 dentitions from the Archaic Period (8000 BC – 1000 AD) from the Lower Pecos region of south-west Texas. They found that the molar tooth loss was very high and speculated that it was caused by a high carbohydrate diet which caused caries and resulted in tooth loss. The anterior teeth were worn rapidly by the fibrous nature of the diet contaminated by grit. These teeth were also used as tools in basket making and weaving and resulted in the early loss of these teeth through wear and damage. Kelley *et al.*, (1991) identified heavy tooth wear as the main cause of antemortem tooth loss.

- Teeth may have been lost because of periodontal disease. The aetiology results from bacteria in the plaque causing an inflammatory response in the periodontal ligament. This effect of this is the loss of the collagen of the periodontal membrane followed by bone resorption. The tooth is no longer supported by bone and becomes mobile and is exfoliated. The alveolus is further resorbed as it no longer supports a tooth. There are several examples of what appears to be tooth loss as the result of localised necrotising periodontitis.
- Extraction of teeth. There is only one example found by the author that could be explained by extraction. This is from the Late Period and shows an absent lower 1st molar (Figure 2b). There is no occlusal wear of the 2nd and 3rd molars and their periodontal condition appears to be excellent. There is no loss of alveolar bone around the site of the missing molar and so it is unlikely that it was lost through trauma or periodontal disease. It is also not congenitally absent as some of the space would have been lost by mesial drift and tilting of the teeth on either side.

Another example of suspected 'extractions' (Craig and Davey, 2001) was thought to be an example of orthodontic extractions in the Graeco-Roman Period. The upper 1st primary molars are missing and there are no permanent 1st premolars in situ which the authors said showed that the unerupted teeth had been enucleated. This seems very unlikely because the remaining 2nd primary molar roots have almost completely resorbed. The 1st primary molars may have exfoliated and the erupting 1st premolars with partially formed roots might have been dislodged during mummification. This was a common cause of postmortem tooth loss. (This had happened to the lower left lateral incisor which was lying dislodged.) Also, the position of the developing canines and 2nd premolars leave insufficient space for a developing 1st premolar which may have been congenitally absent.

These are the only examples found over many years by many investigators. No suitable instruments for extractions have ever been found and there is no record of extraction of teeth in any of the medical papyri. In many skulls there is so little supporting alveolar bone in situ that teeth could have been removed by digital manipulation alone which would have facilitated the drainage of sepsis.

Figures 23a, 23b, 23c and 23d show examples of antemortem tooth loss.

7.7 Wear in 1st, 2nd and 3rd mandibular molars

As was considered in sections 4.11 the condition which has been called 'attrition' in previous research is now known as 'wear' (Soames & Southam, 1993). This is because there is a combination of attrition and abrasion and not merely attrition *per se.* The diet was fibrous and there was the wear of opposing occlusal surface but there also was the addition of contaminants in the preparation of bread which caused abrasion. (Section 2.5.2.) Attrition alone, however, was observed on the interstitial surfaces between adjacent teeth because of the movement of the individual teeth within the periodontal membrane caused by the fibrous nature of the diet.

Begg [2] (1954) had stated that dental 'attrition' was considered necessary to establish an ideal occlusion but Seward (1976) concluded that, as he had established a relationship between wear and temporomandibular osteoarthritis and as osteoarthritis is pathological, therefore wear must also be pathological. This view is diametrically opposed to Begg's hypothesis.

Miles (1963) used the gradient of attrition between the 1st, 2nd and 3rd molar as a means of ageing the fully mature dentition and this was an aid in the ageing of the skulls in this survey. The results demonstrated a consistent reduction in wear from the 1st to the 3rd molar in the majority of the dentitions. In the earlier periods, on occasion the wear was so severe that the laying down of the secondary dentine was overtaken by the wear resulting in the exposure of the pulp and periapical involvement. The lightest wear in all teeth was during the Graeco Roman period. Many of the dentitions from that period looked similar to modern dentition. Figures 24a, 24b, 24c and 24d.

7.8 Condylar Measurement and gender corroboration

Whittaker *et al.,* (1990) had confirmed the sexual dimorphism of the condyle when he had examined 99 adult skulls dating from the 18th Century from Spitalfields in London. 89 of the coffins had breastplates with data about their occupants' age and gender. He found that the mean mesio-lateral measurement for the male was 20.5mm and for the female was 18.5mm. He found a significant difference in the mesio-lateral measurement of male and female condyles. ($P = 0.001$)

In this study of the Egyptian skulls the right mandibular condyle was measured when it was intact but if it was damaged the left condyle was calculated as Whittaker *et al.,* (1985) had found no significant difference between sizes on either side. Vernier calipers were placed on what appeared to be the largest mesio-lateral dimension and this was noted. This measurement was then used to aid corroboration of the assessment of gender made by the skeletal landmarks in this survey. The mean dimension for both genders varied from a low of 19.35 mm in both the Predynastic and the Graeco Roman Periods to a high of 20.56mm in the New Kingdom. When the genders were looked at individually the mean for males was 20.94mm and the mean for females was 18.92mm. There was substantial agreement when it was assumed that females have a measurement of <18.5mm and males > 20.5mm. Wedel *et al.,* (1978) considered that an increase in the mesio-lateral measurement was because of increased functional demand.

Oberg *et al.,* (1971) examined 115 temporomandibular joints obtained during autopsy but, although they were measured, he did not discuss sexual dimorphism.

7.9 Form + surface assessment of the condyle as an index of osteoarthritis

Blackwood (1963) described osteoarthritis as a disease of ageing. He found that it was rare before the 5th decade of life but it was evident much earlier in the Ancient Egyptian. Moffett *et al.,* (1964) believed that the amount of remodelling in the condyle related to missing teeth. Wedel (1971) found no statistical relationship between tooth loss and osteoarthritis. Seward (1976) established a relationship between temporomandibular joint disturbances and attrition. Richards and Brown (1981) found that without associated tooth wear, age alone did not determine osteoarthritis. Eversole *et al.,* (1985) could detect no significant relationship with either posterior or anterior tooth wear and condylar degenerative changes. Hodges (1991) found that age was associated, unquestionably, with osteoarthritis, wear and antemortem tooth loss but she did not find that there was any correlation between osteoarthritis and gender difference or antemortem tooth loss. She established a relationship with age and wear. The possible causes of the osteoarthritis studied in this survey will be analysed further (vide infra).

Waldron (2001) wrote that osteoarthritis can be diagnosed from any two features including osteophtyte formation, pitting of the joint surface and alteration in the bony contour. These were the criteria used in Wedel's (1971) scale and were utilised for this study.

Bone changes are seen in:

Figure 25a which shows osteophyte formation,
Figure 25b demonstrates a large erosion,
Figure 25c which exhibits severe pitting and loss of contour, and Figure 25 d & e illustrate pitting in a condyle with its associated glenoid fossa.

7.10 Caries

Ruffer (1920) said that after his examination of the skulls and the crania from many sources from ancient Egypt '*All this material proved most useful for forming an opinion as to the nature of the dental and another lesions, but owing to the large number of teeth lost in the graves, it was practically useless for a statistical study regarding the incidence of any particular form of dental disease.*' Therefore, when interpreting the statistics there are various factors to be considered. Examining the individual teeth present for caries and calculating the percentage affected has been the method most often used in the past. In this study, the caries incidence in each dentition was used. When the periodontal membrane disintegrates the anterior teeth exfoliate easily because of their single, conical roots. However, they are also less susceptible to caries than are the molars because of the morphology of the crowns. Therefore, if the percentage of caries is taken for a collection of teeth, an inaccurately high rate might be reached.

The teeth lost antemortem may have been carious and exfoliated because of sepsis and this may be of significance. However if the remaining dentition appears to be caries free it may be reasonable to assume that the antemortem tooth loss was not caused by caries but by periodontal disease.

The different sites of carious attack during successive time periods are noteworthy. In the Predynastic and Archaic Periods the sites of carious attack were mainly on the root surface below the amelo-cemental junction. This explains the conclusions drawn about the significance of a glutinous, high-carbohydrate diet from the cereal crops and the preponderance of bread in the diet in the absence of evidence of any oral hygiene regimen at that time.

In the Late Period there is a change in the site of the caries attack with more occlusal and interstitial carious cavities being observed. By the Graeco-Roman period the caries rate had increased and has the appearance of contemporary disease. The occlusal and interstitial caries often occurred in the young compared with the root caries seen in the Predynastic period. There is evidence of a high intake of honey, ***byty,*** by that period but, earlier, in the Predynastic era, honey was used only for religious reasons. Honey was considered to be '*the tears of Ra*' - and the use of honey was a royal prerogative. It was gathered from wild bees and so was very expensive. It was later issued as rations for important officials. By the Middle Kingdom it was found in private tombs as an offering to the gods. In Tutankhamun's tomb there were two jars labelled '*honey of good quality*' (Darby *et al.,* 1977). In later periods it became more readily available. Fruit juice may have been used by the population who were unable to afford honey. There is an illustration in the Middle Kingdom of a liquid being strained through a cloth after having been heated over a fire which is thought to be a concentrated fruit juice (Darby *et al.,* 1977).

Food was also sweetened with dates and figs. Dates were used in the manufacture of beer. Fig seeds were a constituent in a loaf of bread which had a reddish colour, discovered in a tomb. These fruits and concentrated juices are high in sucrose and are very sticky and so would be a cause of an increase in caries in the later periods.

In cases of severe wear occlusal and interstitial caries were prevented from developing. The wear of the occlusal surface eroded the fissures which are the usual sites of carious attack and the attrition of the interstitial surfaces caused the development of contact surfaces from the contact points thereby reducing stagnation areas. However when there was little wear of the teeth these sites were susceptible to caries.

Figures 26a and 26b illustrate root caries,
26c, occlusal caries and
26d and 26e interstitial caries.

7.11 Periodontal breakdown

As the surface of the alveolus is visible, unobscured by desiccated soft tissue there is an advantage in the examination of skeletal over mummified material. Hillson (1996) viewed the aetiology of periodontal disease as multi-factorial and involving inheritance, environment, diet and hygiene. Hildebolt & Molnar (1991) considered that there is little consensus about the aetiology of periodontal disease and were of the opinion that the signs of periodontal disease such as loss of alveolar bone, pitting in the alveolar crest and involvement at the bifurcation of the roots can be identified and separated from fenestrations, dehiscences and bone loss caused by pulpal pathology. These lesions can all be seen without radiology. Clarke *et al.,* 1986) thought that the interpretation of both the severity and occurrence of periodontal disease in dry skulls had often been overestimated. They found that there was a low incidence of the disease which was attributable to gingival inflammation. Despite the absence of known oral hygiene practice, their study of ancient material showed no apparent loss of bone from periodontal origin in over 90% of the cases. In previous work, an increase over 2mm in the distance of the cemento-enamel junction from the alveolar crest had been considered to be alveolar loss in periodontal disease. Clarke and Hirsch (1991) considered that there is an overestimation of the prevalence of periodontal disease as there are two physiological processes which cause an increase greater than 2mm. Firstly, there is the continuous eruption of the tooth which maintains the lower face height when the occlusal tooth wear alone would result in a decrease. Sometimes this is not enough to compensate and this may result in a combined movement of tooth and secondly bone, by the continual growth of cranial structures. This is the *'raison d'etre'* for the measurement of the clinical crown from the alveolar crest to the occlusal surface and for the measurement of the alveolar crest to the amelo-cemental junction and the ratio between them. Brothwell (1994) quoted his survey of 1959 in which he found 74% of Neolithic to Saxon skulls affected by periodontal disease. Hildebolt and Molnar (1991) discussed juvenile periodontitis which is extremely rare occurring in 0.06-0.2% of the population. 1st molars and incisors are the teeth most often affected. They also pointed out that pulpal pathology might mimic periodontal disease. When there is necrosis of the pulp caused by caries, wear or trauma, an apical abscess, cyst or granuloma may develop. The resulting bone loss may be controlled by the immune system or the infection may drain through the crown or a fistula in the alveolus and the bone loss is self limiting. However, sometimes, the infection will drain through the periodontal membrane and the resulting bone loss is so extensive that the tooth exfoliates and the adjacent teeth may be affected.

Examples of three types of periodontal disease were found in this survey of Ancient Egyptian skulls.

Calculus has been discounted in this study because it is difficult to distinguish it from the physico-chemical deposition that can be formed postmortem (Figures 27a and b). Wells (1967) pointed out that postmortem precipitated carbonates are common and they may stain brown from ferruginous infiltration or green from copper salts. These deposits might look like subgingival calculus but their distribution may be atypical.

Alexandersen (1967) wrote, that in early skeletal remains, calculus was found on most of the teeth that he had examined and was most likely to have formed where self cleansing was difficult or impossible because of temporomandibular problems or osteoarthritis. His evidence is a photograph of a mandible with subgingival calculus and its accompanying radiograph which has periodontal thickening. This thickening is more likely to have been caused by pulp necrosis from the caries than as a result of the calculus.

7.11.1 Horizontal periodontal breakdown

For this study it was important, where possible, to separate the horizontal periodontal breakdown from the increase in clinical crown height which is the result of continuous eruption. In the latter the level of root exposure is consistent for the complete dentition and the alveolar crest is covered with dense cortical bone (Murphy 1959, Whittaker *et al.,* 1982, 1985 and 1990, Levers & Darling 1983 and Mays *et al.,* 1995).

When it was considered that there was periodontal breakdown, only a few adjacent teeth were involved. The alveolar crest appeared to be pitted as the result of osteoclast activity. Clarke *et al.* (1986) stated that moderate periodontal disease revealed altered bone morphology of the crest and porosity of the bone surface. They also found that generalised horizontal periodontitis is unusual and is not responsible for tooth loss. Because the skulls were not examined radiologically its presence in this study was judged subjectively. Those which had any evidence of what appeared as horizontal breakdown were noted. No significant difference was found in any group. There were thought to be about 20-30% of skulls involved within each group but in the New Kingdom this figure increased to 38%.

Figures 28a and b).

7.11.2 Vertical periodontal breakdown

This condition is localized around individual teeth. The alveolar border is irregular and there may be interalveolar defects (Figures 29a and b).

The teeth most often affected are the 1st and 2nd molars. When most of the alveolar bone is lost the teeth are exfoliated (Hillson, 1996). There are a variety of aetiologies for this condition. In Ruffer's 1920 survey he considered that occlusal wear was related to periodontal disease. Clarke *et al.*, (1986) concluded that the probable cause was pulpal disease, probably caused by severe wear. Inflammatory agents from the pulp pass through the dentinal tubules to the periodontal ligament and may cause inflammation resulting in resorption.

This survey showed a significant difference between the groups in what was thought to be vertical periodontal disease. The lowest incidence was in the Predynastic group and the highest in the New Kingdom. These results could only be confirmed by radiology to exclude bone loss caused by pulpal disease.

7.11.3 Localised aggressive periodontitis

This condition was found in one skull from the Predynastic period in which the lower left mandibular 1st molar had exfoliated because of bone loss. The periodontal condition in the surrounding and opposing teeth appeared good. There was one similar example from the Old Kingdom. The lower right 1st molar had exfoliated and there was considerable bone loss at the site.

As there is no record of extractions having been performed it provides an explanation for the loss of these teeth. Localised aggressive periodontitis is characterised by the site of attack and its circumpubertal onset. This usually causes the loss of the 1st molar or an incisor with a noticeable loss of alveolus. The rest of the periodontal condition appears to be good (Figures 30a and b). The other features of the disease are impossible to confirm with this material. It may, however, be possible in the future, because of the development of DNA profiling, to test for kinship which may be of relevance as Hodge *et al.*, (2000) have discovered genetic links in families.

7.12 Damage to the alveolar bone attributed to apical or periodontal infection

Damage from what is considered to be a periapical abscess is usually easy to diagnose because of the fistula from the apex to the alveolus or the palate. The tooth affected may show evidence of extreme wear, on occasion with the exposure of the pulp where the secondary dentine has been penetrated, or where there is a carious cavity. What may appear as damage from periodontal abscesses may also be the result of pulp death. Therefore, they have not been differentiated within these statistics.

Figures 31a,b,c,d and e.

7.13 Damage to the alveolar bone attributed to the presence of cysts

Only 18 cysts were recorded in total in the 524 skulls examined. 12 were dental in origin (Figures 32a, 32b, 32c and 32d) but there were 6 which were considered to be developmental and were thought to be nasopalatine.

7.14 Anomalies

Various anomalies were found during this study. These have not been separated in the statistics.

7.14.1 Hypodontia

In Brabant's (1967) study of Neolithic skulls, he found the incidence of hypodontia to be 27%. This comprised of 26% of absent 3rd molars, 0.1-0.8% of lateral incisors and 0.5-1% of premolars. An example seen is shown as Figure 33a.

7.14.2 Hyperdontia

Brabant (1967) found a 0.01-0.05% incidence in Neolithic skulls increasing to 3% in contemporary society. An example from this study is shown as Figure 33b.

7.14.3 Taurodontism

Brabant (1967) said that this condition was usually found in the upper 2nd molar. In the only case that was discovered amongst the Egyptian skulls this was also the case (Figure 34a and b).

7.14.4 Hypercementosis

Brothwell (1963) found this condition in Neanderthal man. It is likely to be the result of extreme occlusal wear. One case was detected in the skulls examined but there might have been more which would have been discovered if radiology had been used (Figure 35).

7.15 Excerebration through the ethmoid bone

In the New Kingdom there were further advances made in mummification techniques. Excerebration was practised and the brain was removed through the cribriform plate of the ethmoid bone, usually via the left nostril. It was noted in many of the specimens in this survey that the entrance hole was often larger than the cribriform plate and the turbinate bones and nasal septum was also destroyed as in Figure 36a. Sometimes the brain was extracted through the orbit and the Pharaoh Ahmose, the first king of the New Kingdom, had his brain removed through the foramen magnum but no examples of this were observed in this survey. The reason that the brain was removed with the viscera was that each experienced rapid decomposition but, unlike the viscera, the brain was not preserved. The ancient Egyptian did not recognise the brain's function (Estes, 1993). Following removal of the brain the cranium was injected with resin. One skull (Figure 36b) shows the force with which the resin was injected as it solidified in a stream when it reached the occipital bone. There was one example of a hole in the base of the skull (Figure 36c) which may have aided the introduction of packing material. This is similar to an illustration in a paper by Strouhal (1986) of a perforation in the same position which he considered had been made artificially for easy access to the cranial cavity.

The *Edwin Smith Papyrus*, Case 6, stated that '*if thou examinest a man having a gaping wound in his head, penetrating to the bone, smashing his skull, rending open the brain of his skull, thou shouldst palpate his wound. Shouldst thou find that the smash which is in his skull like those corrugations which form in molten copper, and something therein throbbing, fluttering under thy fingers, like the weak place of an infant's crown before it becomes whole----'* (Breasted, 1930). Although the function of the brain had not been understood this observation was apposite. The simile about molten copper is graphic and indicates an awareness of the infantile fontanelle. The fontanelle would have been an easy passage for the removal of the brain but it is not known whether this was used or, indeed, if it was customary to remove the brain during the mummification of infants.

Leek (1969) investigated the technique that the embalmers were reported to have used. Herodotus had written in the 5th Century BC that '*as much as possible of the brain is extracted through the nostrils with an iron hook, and what the hook cannot reach is rinsed out with drugs'*. An example of a hook which has been analysed has been found to be made of copper not iron. Leek used sheeps' heads for experimentation and made an entrance hole through a nostril. He discovered that the brain could only be drawn out in small pieces but that the repeated insertion of the hook caused the traumatised tissue to liquefy. By turning the head face downwards the contents of the skull could then drain out.

The majority of the skulls from the New Kingdom were excerebrated but this practice was rather less evident in the Late Period. This confirms that the popularity of this technique was at its zenith in the New Kingdom and then declined.

7.16 Jaw Relationships (Angle's Classification)

Because no radiographs were taken for this study it was not possible to carry out cephalometric analysis of the skull and so only the dental relationships were considered. It was feasible with 258 from the 524 skulls examined to determine these relationships accurately. The factors which were responsible for the exclusion of many of the skulls from this part of the survey were the number of teeth which had been lost both antemortem and postmortem. It was then impossible to be certain about either the buccal relationship of the molars or the incisor relationship. Some of the specimens had missing or fractured maxillae or mandibles. A small number of mandibles had been mismatched with the crania.

It was more accurate to determine the dental relationships in the skulls than in the mummies because the mandibular condyles could be placed precisely in the glenoid fossae. Often, the oral packing of linen and resin, during mummification, prevented the teeth from being in occlusion. Sometimes the mandible was forced out of the glenoid fossa either anteriorly or posteriorly by the mummification process (Harris *et al.*, 1980).

47.3% skulls exhibited a Class 1 occlusion as is illustrated in Figure 37a which is an anterior view and 37b which demonstrates the molar relationships. There was a small percentage, 16.7% of the skulls, which were Class II. Figure 38a and 38b show examples of Class II. This condition was a feature of the New Kingdom Royal Mummies, especially noteworthy in Queen Tetisheri and her granddaughter Queen Ahmose Nefertari and other members of the 18th Dynasty. It could be argued that there was a genetic predisposition as there were many consanguineous marriages (Harris *et al.,* 1980). Another factor which may have been responsible for the Class II relationship may have been oral habits such as thumb or finger sucking. Childhood is often represented by a statue of a child with fingers in its mouth. An example of this is Seneb with his wife and children. He was an official in the Old Kingdom.

Figure 39 illustrates the buccal relationship of one of the large percentage with Class III occlusions may have been as a result of the heavy occlusal wear which allowed the mandible to translate forward. However, these proportions are very different to the findings of Begg [3] (1954). He had examined 800 skulls from Aboriginal Stone Age Man and found only 3% of Class III and 13% Class II (Divisions 1-12% and 2-1%). The large majority were Class I. There was one skull which had a severe anterior open bite but had lost all the anterior teeth antemortem (Figure 39b). There were several skulls with bimaxillary protrusions.

In a recent cephalometric study of a Jordanian population by Hamdan & Rock (2001), 65 adolescents who had an Angle's Class 1 incisor relationship were selected. They were found, in comparison to British norms, to have a reduction in lower face height, with proclined upper and lower incisors and an increased inter-incisal angle, a bimaxillary protrusion.

There were few examples of crowding in evidence. Most arches were well aligned and the interstitial attrition would have aided this (Begg 1977). The few examples were mainly exhibited imbrication of lower incisors and, on occasion a mesio-angular impaction of a mandibular molar was noted (Figures 40 a, b and c).

7.17 Hair Analysis

One very surprising results of the hair analysis (6.19) was the discovery, after 30 minutes of running the chromatogram, that there were peaks which were thought to be the long chain esters (C_{20}) of long chain fatty acids (C_{16}) which are characteristic of beeswax. Manniche (1993) reported that beeswax was used for the care of the hair and jars of beeswax and resin have been found in tombs. Beeswax may have been applied to the hair sample that was analysed. This strand of hair, when examined by electron microscopy, showed, at the highest magnification, what appeared to be a grain of pollen. Another reason for there being beeswax on the hair could have been that it had been used to seal the nose, eyes and mouth and also to model a funerary mask (Brewer *et al.,* 1994 [2]).

The microscopic examination also revealed the scales of the hair on its shaft but the section of hair closest to the flake of scalp was without scales and this would have been the section which would have grown postmortem. Also, in excellent detail are crystals which have been identified as small 'desert roses'. These are composed of hydrous calcium sulphate (gypsum) which occurs in anhydrite form in sedimentary foundations. Gypsum is formed from the hydration of the anhydrite when there are surface waters and ground waters. In arid areas gypsum is deposited as the surface water dries out. This drying out is found in areas such as the Middle East and North Africa (Encyclopaedia Britannica, 2002). If the Nubian had been buried in a pit grave in the sand, which was the form of burial for those of undistinguished birth in the earlier periods in Egypt, he would have been in close proximity to grains of gypsum which in optimum conditions would have crystallised.

Neither the organic nor inorganic profile indicated malnutrition, and so, in this individual, protein energy malnutrition was not the cause of the periodontal condition.

The scanning electron microscopy of the hair is illustrated in Figures 41a, b, c and d.

Chapter 8

Conclusions

8.1 The scope of the study

This study was initiated in order to examine whether a relationship could be demonstrated between the development of agriculture and the variation in dental health noted in the skeletal remains from the Neolithic predynastic settlements from 4000 BC until the end of the Graeco-Roman period in 300 AD. The source of the palaeopathological specimens was the extensive museum collections of skulls in Cambridge and London which were amassed early last century from archaeological sites from all over Egypt.

The hypotheses which are not influenced directly by dietary and dental changes are considered later. The significant dental and dietary changes observed in this study and the relevant Null Hypotheses are as follows;

8.2 Antemortem tooth loss

As the antemortem tooth loss in this survey was found to be greatest in the New Kingdom period the Null Hypothesis (no 2), that there is no significant difference in antemortem tooth loss between samples from each time period, was unproven. Heavy wear was deemed the most important cause of antemortem tooth loss by Kelley *et al*., (1991) but the heaviest wear has been found to be in the Middle Kingdom and the antemortem loss was significantly less than in the New Kingdom.

8.3 Diet and wear

Mean wear of the occlusal surface of the molars was less during the Graeco Roman period than at any other time. The significance of this can be explained by the changes to the diet which occurred during the later period. Athenaeus (AD 200), in '*Deiiphosophista*' or 'The Gastronomers', wrote that Tryphon of Alexandria, 1st Century BC, mentioned various breads including ones made of white wheat flour (Darby *et al*.,1976, Encyclopaedia Britannica). As the bread was no longer made with flour from emmer wheat and was probably more carefully sieved to remove the husks and inorganic particles, this may partly explain the dramatic reduction in occlusal wear. The absence of root caries and appearance of occlusal and interstitial caries may be an indication of the availability of more sweetening agents and their effect on occlusal fissures which were no longer abraded by wear. The interstitial surfaces of the teeth were also more susceptible to carious attack as there was a similar reduction in interstitial attrition and the teeth had contact points instead of surface contacts as in earlier times.

The Null Hypothesis (no.3) that there is no significant difference in molar wear is not proved.

8.4 Osteoarthritis in the mandibular condyle.

Osteoarthritis was assessed by using Wedel's scale (*vide supra* 5.4) as an index. It was found to be at its maximum during the Archaic and Old Kingdom periods and at its minimum during the Graeco Roman period. Oberg (1971) explained one of the causes of osteoarthritis as antemortem loss of teeth but this was not found to have a direct relationship in the Ancient Egyptian skulls examined.

The Null hypothesis (no.5) that there was no significant difference in degree of osteoarthritis between samples is not proved.

8.5 Dietary influences on caries and periodontal disease

During the Predynastic period, the main cultivated cereal was emmer wheat. This was used for bread which was a staple of the Egyptian diet (*vide supra* 2.5.2). When mixed with saliva during mastication it became very glutinous and so would have adhered to teeth. Previously examined bread from tombs was shown to contain husks and abrasive inorganic particles.

There is no knowledge of any oral hygiene regimen in Egypt in ancient times. Clarke & Hirsch (1991) described a continuum between plaque, gingivitis and periodontitis but stated that poor oral hygiene does not always lead to periodontitis. The incidence of root caries was found to be greatest during the Predynastic period but there was little evidence of interstitial or occlusal caries in this population. In this study 84% of the predynastic skulls studied exhibited no caries. This percentage remained at almost the same level in the skulls from the Old Kingdom and Middle Kingdom. By the New Kingdom and Late Period, the caries free percentage reduced to around 75% and by the Graeco-Roman period it had further decreased to 66.7% and the site of the caries changed from root to occlusal and interstitial caries. This research has suggested that the aetiology of the increase in caries and the change of site was the result of the increased availability of honey and fruit juices which were used for sweetening. There is reference to wild

sugarcane found in the predynastic Omarian deposits (Hayes 1964) but there is no record of the use of sugar in Ancient Egypt. Cariogenic fruit juice which was used as a sweetening agent has been found in a 12th Dynasty (Middle Kingdom) tomb.
Pliny (23-79AD) wrote of 'honey from reeds' which may be sugar cane (Aykroyd, 1967). However sugarcane was brought by the Arab invaders to Egypt for cultivation in the 7th Century AD (Darby *et al.*, 1977) and so can not be implicated in the increase in caries noted in the Graeco-Roman period.

The Null hypothesis (no. 6) stating that there is no significant difference in the caries rate is therefore not proved.

8.6 Periodontal disease and periapical disease

More of the population (43%) during the New Kingdom, in the group examined, appeared to be affected by periodontal disease than during any other time period. They also had a slightly higher incidence of, what appeared to be, fistulae in the alveolar bone, possibly as the result of pulp death. There may be an over assessment of what was considered to be periodontal disease as it is impossible to be certain about the aetiology of some alveolar bone loss. Clarke *et al.*, (1986) found that heavy tooth wear can cause pulp death resulting in inflammatory mediators from the pulp reaching the periodontal ligament via the dentinal tubules which may cause inflammation of the periodontium. This may give the impression of periodontal disease.

The Null hypothesis (no.7) that there is no significant difference in periodontal disease at different time periods is not proved as there is a variation in periodontal disease with the highest incidence having occurred in the New Kingdom.

8.7 Abscesses and cysts

The Null hypothesis (no.8) that there is no significance in the numbers of abscesses between time samples from each time period is proved. The number of cysts was too small to either prove or disprove the Null hypothesis.

8.8 Continuous eruption and the relationship between it and osteoarthritis/age/wear

No significant differences have been found in the measurement of continuous eruption in each of the periods. Clarke *et al.*, (1986) stated that an increase in the distance between the alveolar crest and cemento-enamel junction was physiological and continued throughout life and is not a sign of periodontal disease. In past research it had been erroneously assumed that any increase in the distance between the amelo-cemental junction and the alveolar crest >2mm was a sign of periodontal disease (Clarke and Hirsch, 1991) which led to an overestimation. However, in this study, significant associations have been found in the relationship between the ratio of continuous eruption to crown height and wear in the 1st molar. It might therefore be concluded that, the greater the wear there is in the 1st molar, the greater the stimulation for continuous eruption. There was also a relationship between the ratio of continuous eruption to crown height and osteoarthritis of the mandibular condyle. Whether the ratio of the continuous eruption is increased by the degree of osteoarthritis, or the reverse is true, is uncertain. The Null hypothesis (no.9) is not proven.

This research has also shown that a direct relationship has been found between:

- age and wear in the 1st molar,
- age and osteoarthritis of the temporomandibular condyle
- a significant relationship between wear in the 1st molar and osteoarthritis.

The Null hypothesis (no.10) is not proven.

8.9 Hair and dietary intake

Because it was considered that the Ancient Egyptian could have suffered from protein energy malnutrition because they had a high cereal diet and experienced periods of famine, some red hairs taken from the skull of an Old Kingdom Nubian were examined by Mass Spectrometry. Red hairs had been selected as they might indicate the 'flag sign' of kwashiorkor. However this investigation established that the subject had taken a high protein dietary intake. The red hair in this specimen may have been naturally occurring or could have lost pigment by contact with the hot sand.
The Null Hypothesis (no.13) is proved for this subject (*vide infra*).

8.10 Gender and age at death

The Null Hypothesis (no.1) that there was no significant difference in gender and age at death between samples from each time period was proved for gender but not proved for age as there was an increase in child and adolescent deaths in the Graeco-Roman period. There is no obvious explanation for this and viral epidemics may have been responsible. It is also possible that by that time the burial of infants and children, in a more formal way, was common practice. This would have influenced this statistic.

8.11 The gender and condylar size

There was substantial agreement with the gender classification from the skull when compared with the assessed gender of the larger (male) and the smaller (female) condyles (Whittaker,1990). Therefore the Null Hypothesis (no. 4), that there was no difference in condylar size and assessed gender, is not proved.

8.12 Anomalies

The Null Hypothesis (no. 11) that there was no difference in anomalies found was disproved as there was a significant increase in the anomalies present in the dentitions from the Late Period.

8.13 Excerebration through the ethmoid bone

This was not common practice until the New Kingdom and was not noted before that period in any of the skulls examined. The incidence at that period was 55.1% and this reduced dramatically to 16.7% in the Late Period with a further reduction to 13.5% in the Graeco-Roman Period. Because of damage to many of the crania the percentage may have been higher than recorded. The Null Hypothesis, that there were no significant differences, was thus disproved.

8.14 Radiology

One significant tool, which could have authenticated the observed diagnoses of caries and periodontal involvement, would have been radiological investigation. This technique was not used because of logistical difficulties, safety and cost. Some skulls would have benefited from a further investigation as some caries may have been undetected and periapical lesions, in the early stages, will not have been observed.

Notwithstanding, this research has validity, as it has given an overall picture of the changing dental health of the Ancient Egyptian over 4 millennia and, despite an apparently healthy diet, the suffering that was caused by its effect on the individual.

RECOMMENDATIONS FOR FURTHER STUDY

Further work investigating the nutritional health of ancient Egyptian remains by hair analysis using mass spectrometry might give additional insight into the diet. Most of the mummified heads have hair available and, as only a few hairs are necessary for this investigation, it would be initially non-destructive. Only, very occasionally, are hairs available on a dry skull and such an example provided the source for this study.

It also would be worthwhile for an assessment of dental health in collections of skulls from similar time periods but from differing climatic zones to be made to evaluate the effect of differing diets on the dentition.

In view of the lack of evidence of dental treatment in ancient Egypt, it would be of interest to examine early medical and dental writings and dental artefacts from other societies such as Ayuravedic medicine and the examples of Phoenician and Etruscan dentistry.

Table 6: ***Estimated age at death in different time periods.***

Age	*Predynastic*	*Archaic Old Kingdom*	*Middle Kingdom*	*New Kingdom*	*Late Period*	*Ptolemaic & Roman*	*Total*
1-10 yrs		*3 2.5%*		*3 3.5%*	*3 5.3%*	*4 5.6%*	*13 2.6%*
10-20 yrs	*13 16.3%*	*13 11.0%*	*5 5.6%*	*13 15.1%*	*5 8.8%*	*13 18.3%*	*62 12.4%*
20-30 yrs	*36 45%*	*46 39.0%*	*26 29.2%*	*20 23.3%*	*15 26.3%*	*27 38.0%*	*170 33.9%*
30-40 yrs	*17 21.3%*	*32 27.1%*	*31 34.8%*	*30 34.9%*	*21 36.8%*	*17 23.9%*	*148 29.5%*
Elderly	*14 17.5%*	*24 20.3%*	*27 30.3%*	*20 23.3%*	*13 22.8%*	*10 14.1%*	*108 21.6%*
Total	*80 100%*	*118 100%*	*89 100%*	*86 100%*	*57 100%*	*71 100%*	*501 100%*

Table 7: ***Gender identification.***

Time period	*Female*	*Male*	*Adoles.*	*Child*	*Uncertain*	*Total*
Predynastic	*37 17.5%*	*43 16.2%*			*10 52.6%*	*90 17.2%*
Archaic & Old Kingdom	*55 26.1%*	*60 22.6%*	*1 6.7%*	*3 21.4%*	*1 5.3%*	*120 22.9%*
Middle Kingdom	*43 20.4%*	*42 15.8%*	*2 13.3%*		*3 15.8%*	*90 17.2%*
New Kingdom	*27 12.8%*	*53 20%*	*3 20.0%*	*4 28.6%*	*2 10.5%*	*89 17.0%*
Late Period	*25 11.8%*	*29 10.9%*	*2 13.3%*	*3 21.4%*	*1 5.3%*	*60 11.5%*
Ptolemaic & Roman	*24 11.4%*	*38 14.3%*	*7 46.7%*	*4 26.6%*	*2 10.5%*	*75 14.3%*
Total	*211 100%*	*265 100%*	*15 100%*	*14 100%*	*19 100%*	*524 !00%*

Table 8: ***Validity of gender using museum records.***

MUSEUM

OBSERVER

	male	*female*
Male	*1*	*0*
Female	*2*	*6*
Uncertain	*0*	*1*

Table 9: ***Antemortem tooth loss per dentition within each time period.***

Group	*Mean*	*Std. deviation*	*Minimum*	*Maximum*
Predynastic	*1.76*	*3.77*	*0*	*23*
Archaic & Old Kingdom	*2.29*	*4.60*	*0*	*25*
Middle Kingdom	*2.13*	*3.30*	*0*	*15*
New Kingdom	*4.26*	*6.43*	*0*	*32*
Late Period	*4.02*	*5.41*	*0*	*29*
Ptolemaic & Roman	*2.76*	*4.77*	*0*	*23*

Table 10: ***Significant differences in antemortem tooth loss within each time period.*** **(P<0.05)**

	Archaic& Old Kingdom	*Middle Kingdom*	*New Kingdom*	*Late Period*	*Ptolemaic & Roman*
Predynastic			*		
Archaic&OK					
Mid. Kingdom			*		
New Kingdom					
Late Period					
Ptol. & Roman					

ANOVA test: Significant differences between each 2 groups (P<0.05) are indicated by *

Table 11: ***Mean wear in 1st molar in different time period.***

Group	*Mean*	*Std. deviation*	*Minimum*	*Maximum*
Predynastic	*4.92*	*1.48*	*2*	*7*
Archaic & Old Kingdom	*4.97*	*1.52*	*1*	*8*
Middle Kingdom	*5.51*	*1.41*	*2*	*8*
New Kingdom	*4.47*	*1.73*	*1*	*7*
Late Period	*5.04*	*1.54*	*1*	*7*
Ptolemaic & Roman	*3.40*	*1.45*	*1*	*7*

Table 12: ***ANOVA test of mean wear in1st molar in different time periods.*** **(P<0.05)**

	Archaic& Old Kingdom	*Middle Kingdom*	*New Kingdom*	*Late Period*	*Ptolemaic & Roman*
Predynastic					*
Archaic & O.K.					*
Mid. Kingdom			*		*
New Kingdom					*
Late Period					*
Ptol. & Roman					

ANOVA test: Significant differences between each 2 groups (P<0.05) are indicated by *

Table 13: ***Crosstabulation of tooth wear in different time periods of 1st molar with mild wear 1-4 and severe wear 5-8.***

	WEAR 1-4 mild	*WEAR 5-8 severe*	*TOTAL*
Predynastic	30 38%	49 62.0%	79 100%
Archaic & Old Kingdom	39 39.4%	60 60.6%	99 100%
Middle Kingdom	25 32.9%	51 67.1%	76 100%
New Kingdom	32 48.5%	34 51.5%	66 100%
Late Period	18 36.7%	31 63.3%	49 100%
Ptolemaic & Roman	44 83.0%	9 17.0%	53 100%
Total	188 44.5%	234 55.5%	422 100%

Table 14: ***Mean wear in 1st molar in different time periods.***

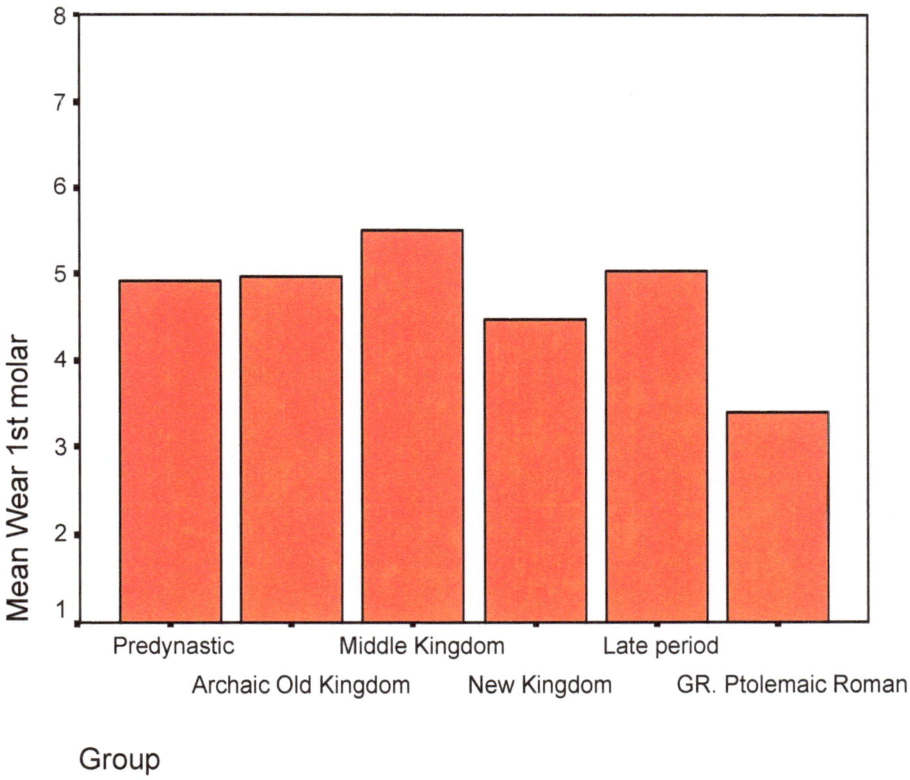

Table 15: ***Mean wear in 2nd molar in different time periods.***

Group	*Mean*	*Std. deviation*	*Minimum*	*Maximum*
Predynastic	3.60	1.58	1	7
Archaic & Old Kingdom	3.48	1.51	1	7
Middle Kingdom	4.14	1.79	1	7
New Kingdom	3.43	1.72	1	7
Late Period	3.79	1.44	1	6
Ptolemaic & Roman	2.43	1.36	1	6

Table 16: ***Anova test of mean wear in 2nd molar in different time periods.***

	Archaic& Old Kingdom	*Middle Kingdom*	*New Kingdom*	*Late Period*	*Ptolemaic & Roman*
Predynastic					*
Archaic & OK					*
Mid. Kingdom					*
New Kingdom					*
Late Period					*
Ptol.& Roman					

Table 17: ***Mean wear in 2nd molar in different time periods.***

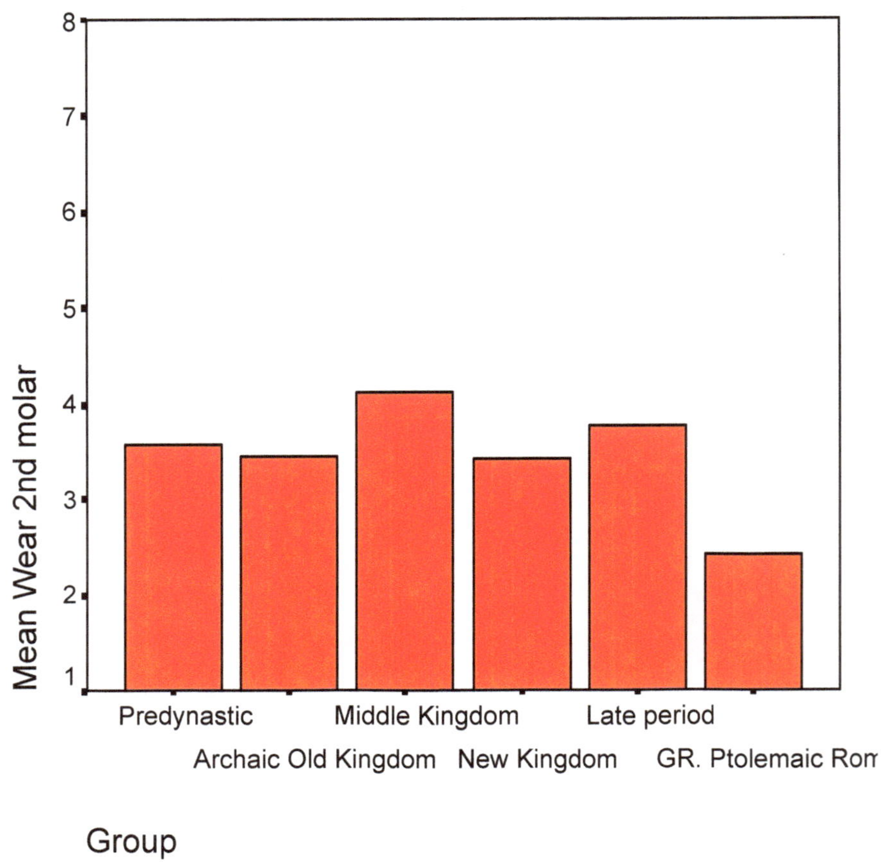

Table 18: *Mean wear in 3rd molar in different time periods.*

Group	Mean	Std. deviation	Minimum	Maximum
Predynastic	2.36	1.37	1	7
Archaic & Old Kingdom	2.33	1.32	1	7
Middle Kingdom	2.80	1.71	1	7
New Kingdom	2.49	1.36	1	7
Late Period	2.71	1.45	1	6
Ptolemaic & Roman	1.92	0.98	1	5

Table 19: *Anova test of mean wear in 3rd molar in different time periods.*

	Archaic& Old Kingdom	Middle Kingdom	New Kingdom	Late Period	Ptolemaic & Roman
Predynastic					
Archaic & OK					
Mid. Kingdom					*
New Kingdom					
Late Period					
Ptol. & Roman					

ANOVA test: Significant differences between each 2 groups (P<0.05) are indicated by *

Table 20: *Mean wear in 3rd molar in different time periods.*

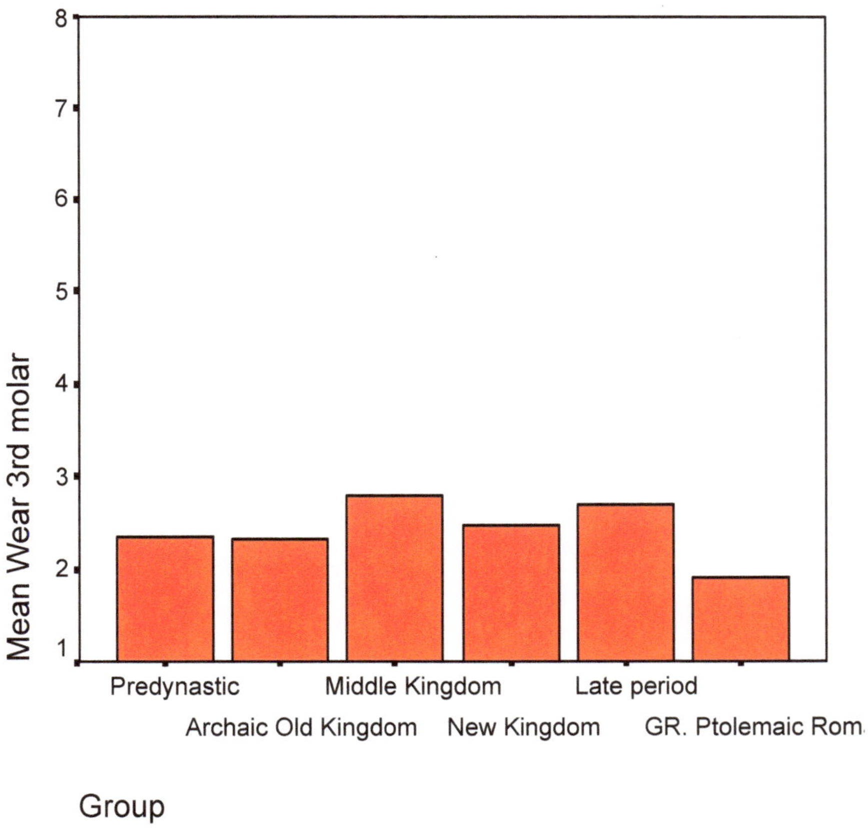

Table 21: ***Mean mesiolateral condylar measurements in all time periods.***

Group	*Mean mesiolateral measurement in mm.*	*Std. deviation*	*Minimum*	*Maximum*
Predynastic	*19.35*	*1.86*	*14.34*	*24.00*
Archaic & OK	*20.06*	*2.43*	*11.81*	*24.32*
Middle Kingdom	*19.69*	*2.39*	*15.27*	*31.23*
New Kingdom	*20.54*	*2.34*	*13.32*	*29.80*
Late Period	*19.74*	*1.88*	*15.76*	*23.98*
Ptol. & Roman	*19.35*	*2.47*	*11.80*	*24.93*

Table 22: ***Anova test of mean condylar measurements in all time periods.***

	Archaic& Old Kingdom	*Middle Kingdom*	*New Kingdom*	*Late Period*	*Ptolemaic & Roman*
Predynastic			*		
Archaic & OK					
Mid Kingdom					
New Kingdom					*
Late Period					
Ptol. & Roman					

ANOVA test: Significant differences between each 2 groups (P<0.05) are indicated by *

Table 23: ***Relationship between gender and condylar measurement.***

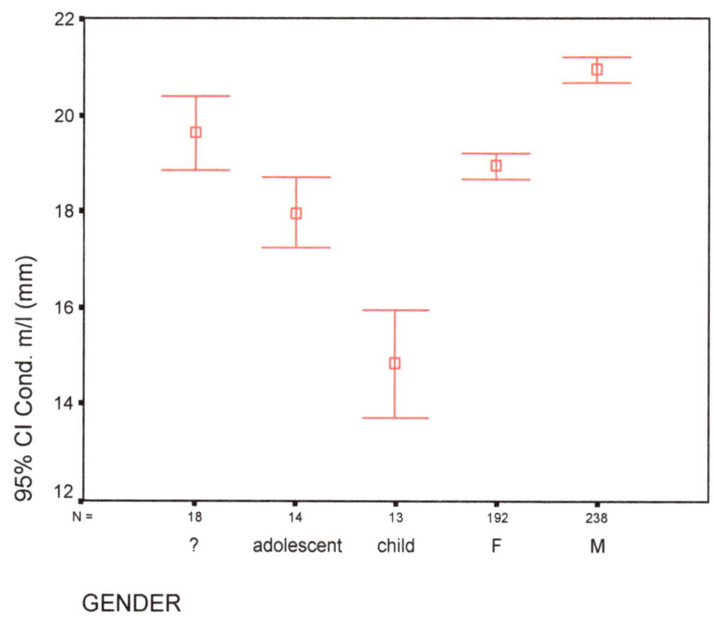

Table 24: ***Relationship between gender and condylar measurement.*** **(t-test, P<0.001)**

Gender by Observation	Mean	Std.deviation	Minimum	Maximum
M. condyle m/l measurement	20.94	2.01	11.81	29.80
F condyle m/l measurement	18.92	1.76	14.34	31.23

Table 25: ***Crosstabulation of gender estimated from observation and condylar measurement.***

Gender by Observation	M condyle > 21	f condyle < 18.5	Total
No. of males	122 83%	25 17%	147 100%
No. of females	17 17.3%	81 82.7%	98 100%
Total	139 56.7%	106 43.3%	245 100%

Table 26: ***Form + surface as a measurement of osteoarthritis.***

Group	Mean form + surface	Std. deviation	Minimum	Maximum
Predynastic	2.32	1.24	0	5
Archaic & Old Kingdom	2.55	1.48	0	6
Middle Kingdom	2.36	1.09	0	5
New Kingdom	2.04	1.26	0	5
Late Period	2.13	1.29	0	5
Ptolemaic & Roman	1.38	1.35	0	5

Table 27: ***ANOVA test of Form plus surface as a measurement of osteoarthritis in different time periods.***

	Archaic& Old Kingdom	Middle Kingdom	New Kingdom	Late Period	Ptolemaic & Roman
Predynastic			*		*
Archaic & OK					*
Mid. Kingdom					*
New Kingdom					*
Late Period					*
Ptol. & Roman					

ANOVA test: Significant differences between each 2 groups (P<0.05) are indicated by *

Table 28: ***Caries: affected teeth present per dentition.***

Group	Mean no. of TEETH AFFECTED PER DENTITION.	Std. deviation	Minimum	Maximum
Predynastic	0.23	.64	0	4
Archaic & Old Kingdom	0.33	.83	0	5
Middle Kingdom	0.34	.85	0	4
New Kingdom	0.35	.71	0	3
Late Period	0.40	.83	0	4
Ptolemaic & Roman	0.71	1.31	0	7

Table 29: ***ANOVA test for Caries: teeth affected per dentition.***

	Archaic& Old Kingdom	*Middle Kingdom*	*New Kingdom*	*Late Period*	*Ptolemaic & Roman*
Predynastic					*
Archaic & OK					*
Mid. Kingdom					
New Kingdom					
Late Period					
Ptol. & Roman					

ANOVA test: Significant differences between each 2 groups (P<0.05) are indicated by *

Table 30: ***Caries crosstabulation of number of teeth affected.***

	0	1	2	3	4	5	7	Total
Predynastic	76 84.4%	9 10.0%	4 4.4%		1 1.1%			90 100%
Archaic & O.K.	98 81.7%	12 10.0%	6 5.0%	2 1.7%	1 0.8%	1 0.8%		120 100%
Middle Kingdom	75 83.3%	4 4.4%	7 7.8%	3 3.3%	1 1.1%			90 100%
New Kingdom	68 76.4%	13 14.6%	6 6.7%	2 2.2%				89 100%
Late Period	45 75.0%	9 15.0%	4 6.7%	1 1.7%	1 1.7%			60 100%
Ptolemaic & Roman	50 66.7%	11 14.7%	7 9.3%	4 5.3%	1 1.3%	1 1.3%	1 1.3%	75 100%
Total	412 78.6%	58 11.1%	34 6.5%	12 2.3%	5 1.0%	2 0.4%	1 0.2%	524 100%

Table 31: ***Horizontal periodontal breakdown in different time periods.*** **(χ^2 test P=0.069).**

Horizontal periodontal breakdown	*0 no*	*1 yes*	*total*
Predynastic	66 73.3%	24 26.7%	90 100%
Archaic & Old Kingdom	91 75.8%	29 24.2%	120 100%
Middle Kingdom	72 80%	18 20%	90 100%
New Kingdom	55 61.8%	34 38.2%	89 100%
Late Period	44 73.3%	16 26.7%	60 100%
Ptolemaic & Roman	60 80.0%	15 20.0%	75 100
Total	388 74%	136 26%	524 100%

Table 32: ***Vertical periodontal breakdown in different time periods.* (χ^2 test P=0.01)**

Vertical periodontal breakdown	*0 no*	*1 yes*	*total*
Predynastic	*71* *78.9%*	*19* *21.1%*	*90* *100%*
Archaic & Old Kingdom	*92* *76.7%*	*28* *23.3%*	*120* *100%*
Middle Kingdom	*62* *68.9%*	*28* *31.1%*	*90* *100%*
New Kingdom	*46* *51.7%*	*43* *48.3%*	*89* *100%*
Late Period	*43* *71.7%*	*17* *28.3%*	*60* *100%*
Ptolemaic & Roman	*56* *74.7%*	*19* *25.3%*	*75* *100%*
Total	*370* *70.6%*	*154* *29.4%*	*524* *100%*

Table 33: ***Vertical periodontal breakdown in different time periods.***

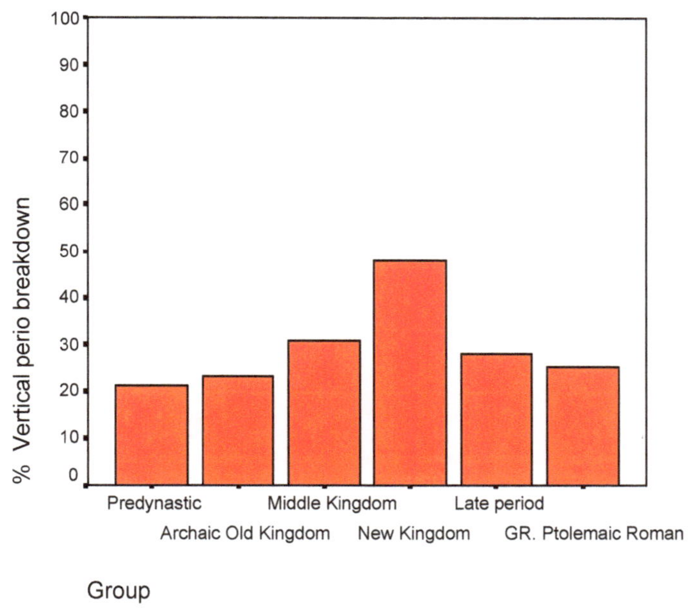

Table 34: ***Alveolar damage probably as a result of apical or periodontal infection.***

group	*mean no. of abscesses PER DENTITION.*	*Std. deviation*	*minimum*	*maximum*
Predynastic	*.71*	*1.26*	*0*	*5*
Archaic & Old Kingdom	*.86*	*1.37*	*0*	*6*
Middle Kingdom	*.96*	*1.55*	*0*	*8*
New Kingdom	*1.08*	*1.63*	*0*	*6*
Late Period	*.58*	*1.05*	*0*	*5*
Ptolemaic & Roman	*.51*	*1.12*	*0*	*6*

Table 35: ***Alveolar damage probably caused by cysts..***

no of cysts	*0*	*1*	*2*	*total*
Predynastic	*89* *98.9%*	*1* *1.1%*		*90* *100%*
Archaic & Old Kingdom	*115* *95.8%*	*4* *3.3%*	*1* *0.8%*	*120* *100%*
Middle Kingdom	*89* *98.9%*	*1* *1.1%*		*90* *100%*
New Kingdom	*87* *97.8%*	*2* *2.2%*		*89* *100%*
Late Period	*53* *88.3%*	*7* *11.7%*		*60* *100%*
Ptolemaic & Roman	*74* *98.7%*	*1* *1.3%*		*75* *100%*
Total	*507* *99.8%*	*16* *3.1%*	*1* *0.2%*	*524* *100%*

Table 36 : ***Anomalies.***

Anomalies in each dentition.	*0*	*1*	*2*	*3*	*total*
Predynastic	*78* *86.7%*	*10* *11.1%*	*2* *2.2%*		*90* *100%*
Archaic & Old Kingdom	*107* *89.2%*	*10* *8.3%*	*3* *2.5%*		*120* *100%*
Middle Kingdom	*76* *84.4%*	*13* *14.4%*	*1* *1.1%*		*90* *100%*
New Kingdom	*77* *86.5%*	*12* *13.5%*			*89* *100%*
Late Period	*38* *63.3%*	*19* *31.7%*	*2* *3.3%*	*1* *1.7%*	*60* *100%*
Ptolemaic & Roman	*62* *82.7%*	*13* *17.3%*			*75* *100%*
Total	*438* *83.6%*	*77* *14.7%*	*8* *1.5%*	*1* *.2%*	*524* *100%*

Categories 1-3 were combined because of small numbers.

Table 37: ***Mismatched mandibles.***

time Period	*no of mismatched mandibles*
Predynastic	*0*
Archaic & Old Kingdom	*5*
Middle Kingdom	*0*
New Kingdom	*1*
Late Period	*4*
Ptolemaic & Roman	*5*

Table 38: ***Excerebration through the ethmoid bone.***

	0 *no excerebration*	1 *yes excerebration*	***Total***
New Kingdom	*40* *44.9%*	*49* *55.1%*	*89* *100%*
Late Period	*25* *83.3%*	*5* *16.7%*	*30* *100%*
Ptolemaic & Roman	*64* *86.5%*	*10* *13.5%*	*74* *100%*
Total	*129* *66.8%*	*64* *32.2%*	*193* *100%*

Table 39: ***Relationship between wear of 1st molar and age.***

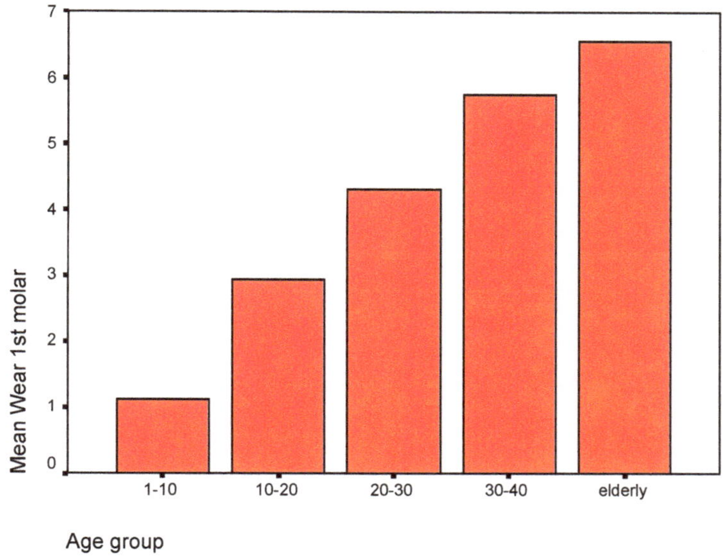

Table 40: ***Relationship between osteoarthritis (Form + Surface) and age.***

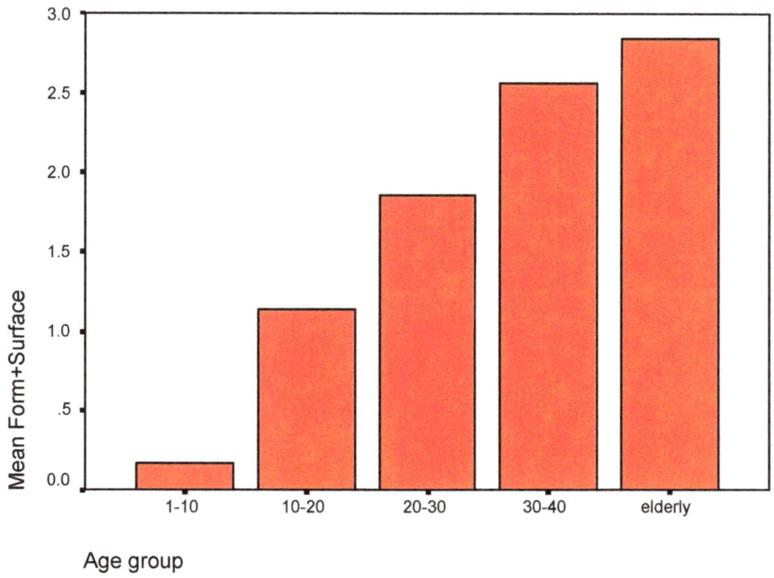

Table 41: ***Relationship between wear in the 1st molar and osteoarthritis of the temporomandibular joint.***

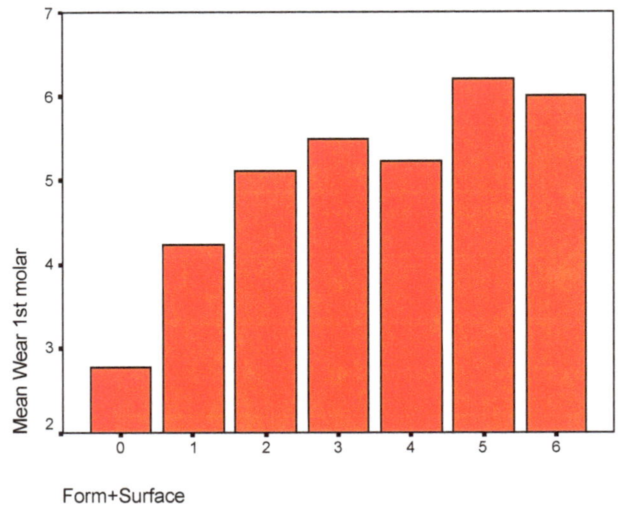

Table 42: ***Relationship between wear in the 1st molar and osteoarthritis of the temporomandibular joint.***

Nonparametric Correlations

Spearman's correlation		*wear 1st molar*	*form + surface.*
Wear 1st molar	*correlation coefficient*	1.000	.448*
	Sig. (2-tailed)	.	.000
	N	422	384
Form + surface	*correlation coefficient*	.448*	1.000
	Sig. (2-tailed)	.000	.
	N	384	477

***Correlation is significant at the .01 level (2-tailed).**

Table 43: ***Measurement of continuous eruption.***

Group	*Mean Cont. eruption in mm*	*Std. deviation*	*Minimum*	*Maximum*
Predynastic	3.04	1.60	1.00	9.77
Archaic & OldKingdom	3.63	1.53	0.00	8.87
Middle Kingdom	3.31	1.14	1.00	8.00
New Kingdom	3.55	1.74	0.00	8.98
Late Period	3.60	1.23	1.40	7.96
Ptolemaic & Roman	3.41	1.24	1.68	7.12

Table 44: ***Measurement of relationship of ratio of the continuous eruption to crown height and the wear in 1st molar.***

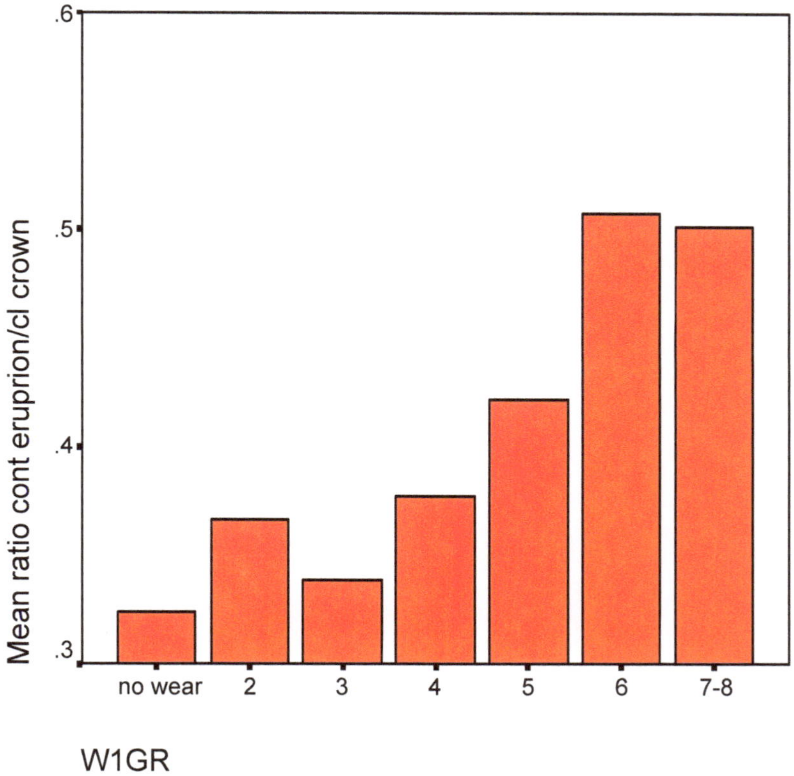

Table 45: ***Measurement of relationship of ratio of the continuous eruption to crown height and Form + Surface of the temporomandibular joint.***

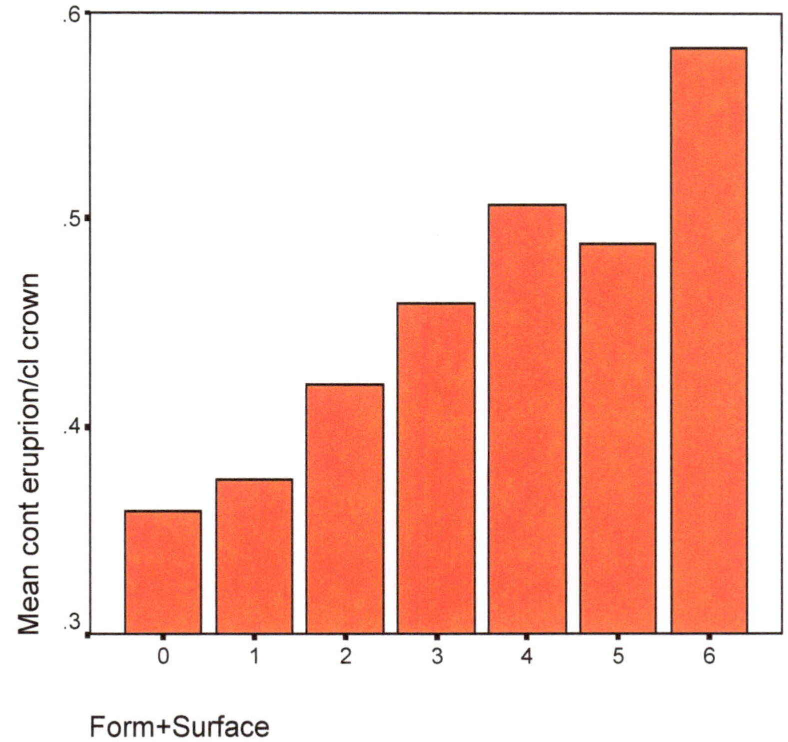

Table 46: Jaw relationships (Angle's Classification).

	Frequency	*Percent*	*Valid Percent*	*Cumulative Percent*
Class 1	*122*	*47.3*	*47.3*	*47.3*
Class 2	*43*	*16.7*	*16.7*	*64.0*
Class 3	*93*	*36.0*	*36.0*	*100.0*
Total	*258*	*100*	*100*	

Table 47: Total Ion Chromatogram (TIC) of the ancient red hair sample.

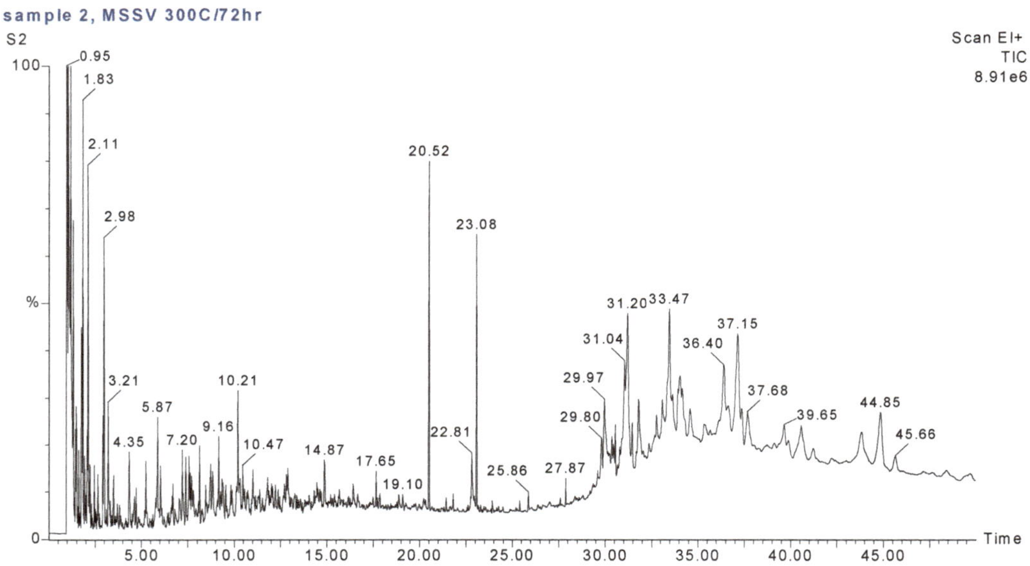

Table 48: Total Ion Chromatogram (TIC) Red hair from Bolton MummyModern red hair

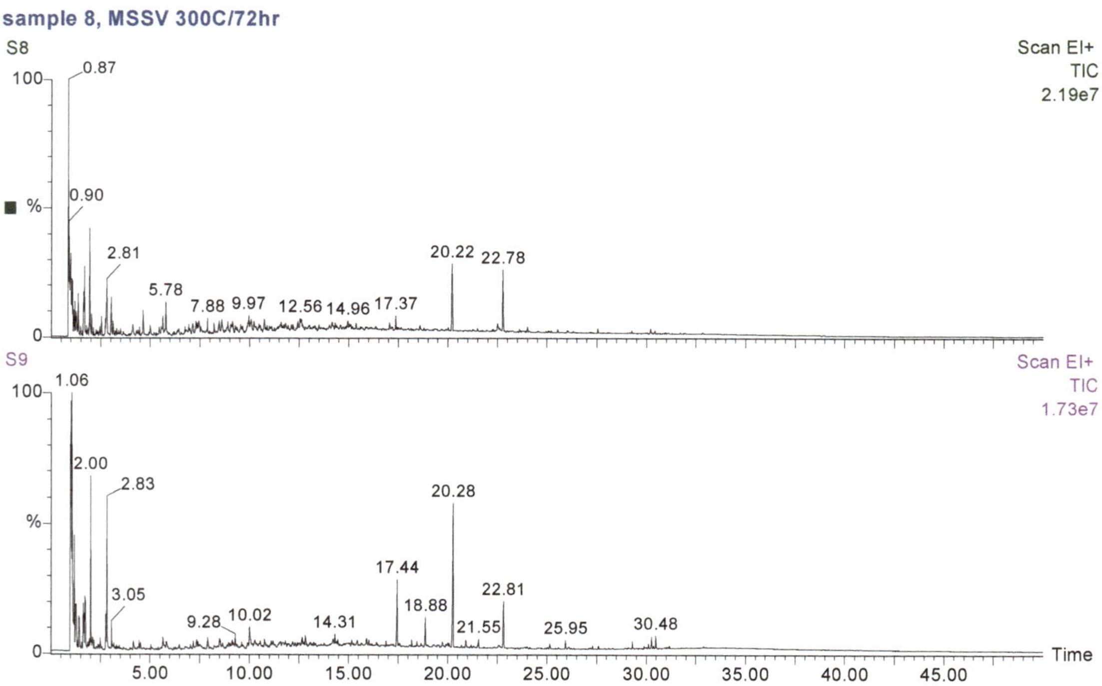

Table 49: Total Ion Chromatogram (TIC) (17-25 minutes)

Nubian hair
Red hair from Bolton Mummy
Modern red hair

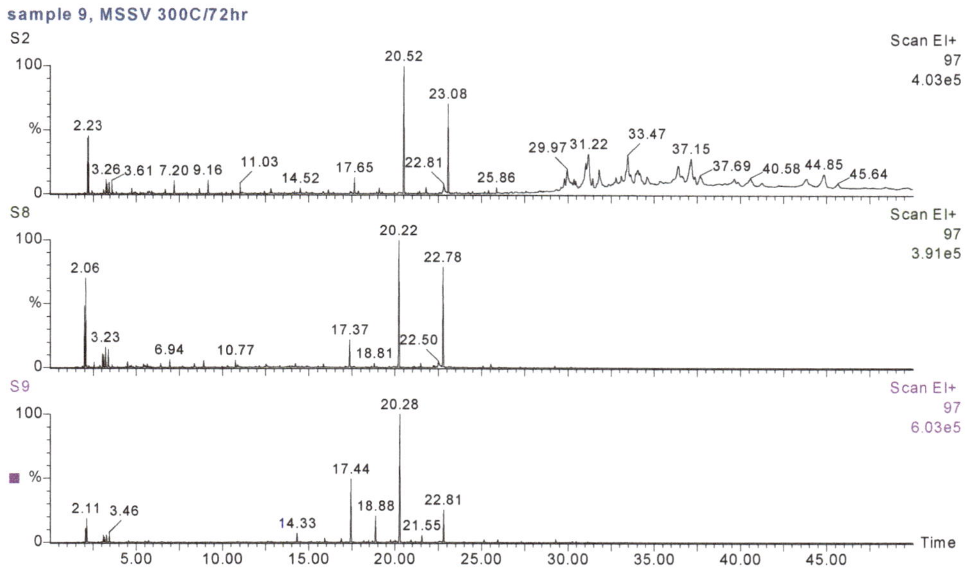

Table 50: Total Ion Chromatogram (TIC) (23 minutes)

Nubian hair
Near match library specimen

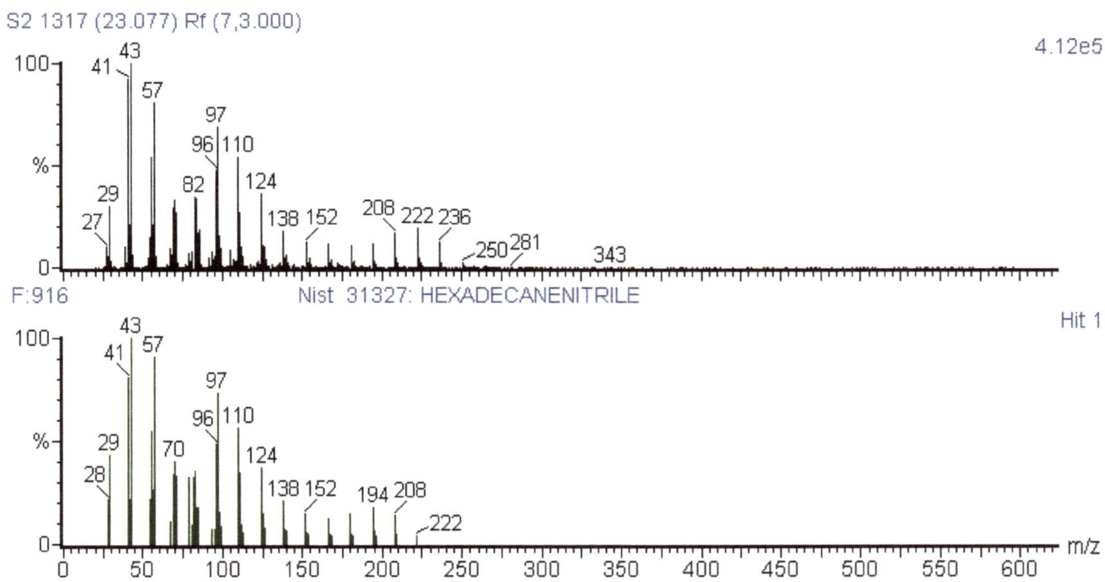

Table 51: Total Ion Chromatogram (TIC) (30 minutes)

Long chain ester characteristic of beeswax
Nubian hair

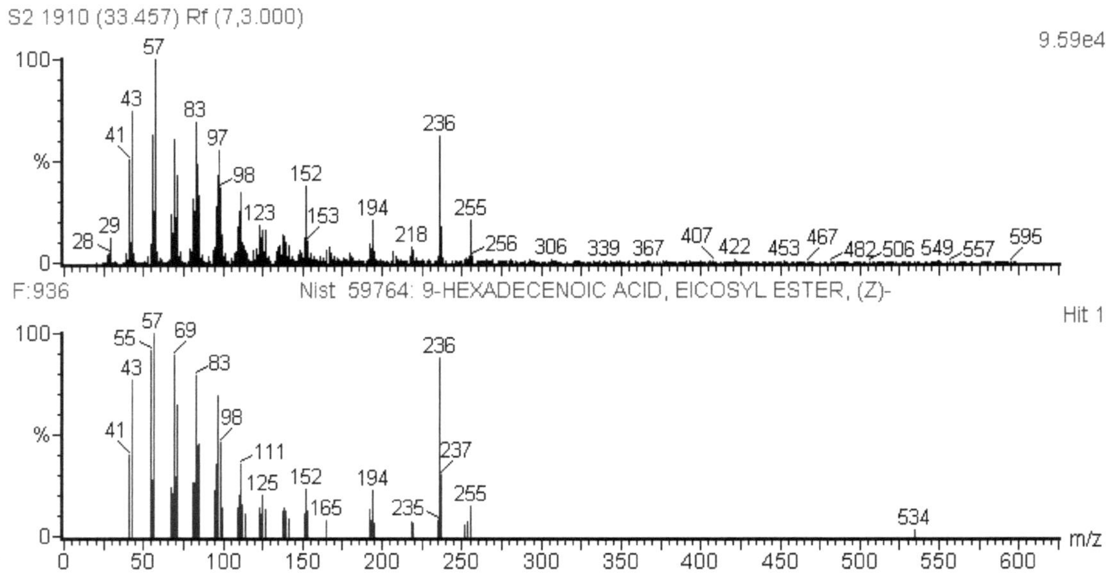

Table 52: Mass Chromatogram (TIC)

Nubian hair chromatogram diagnostic for steroids

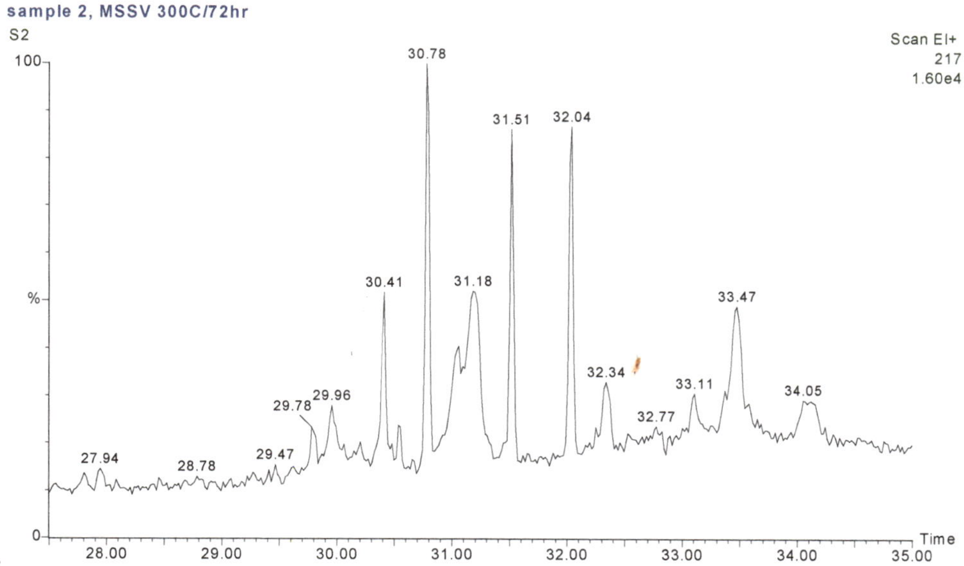

The Nubian hair at 32.04 .
Library match diagnostic for steroids.

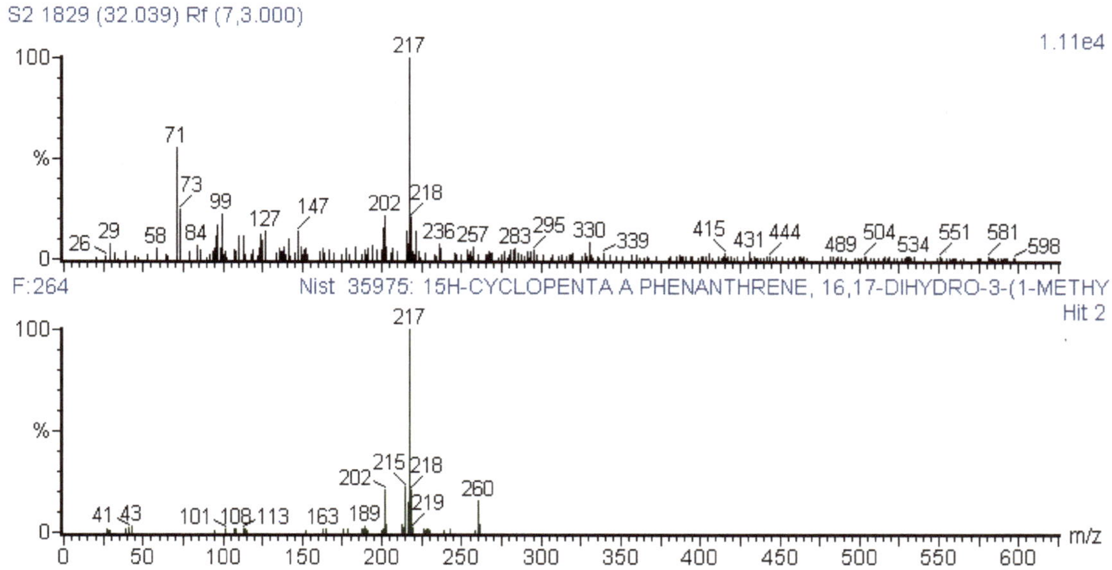

Table 53: Mass Chromatogram (TIC)

Nubian hair chromatogram diagnostic for lipids

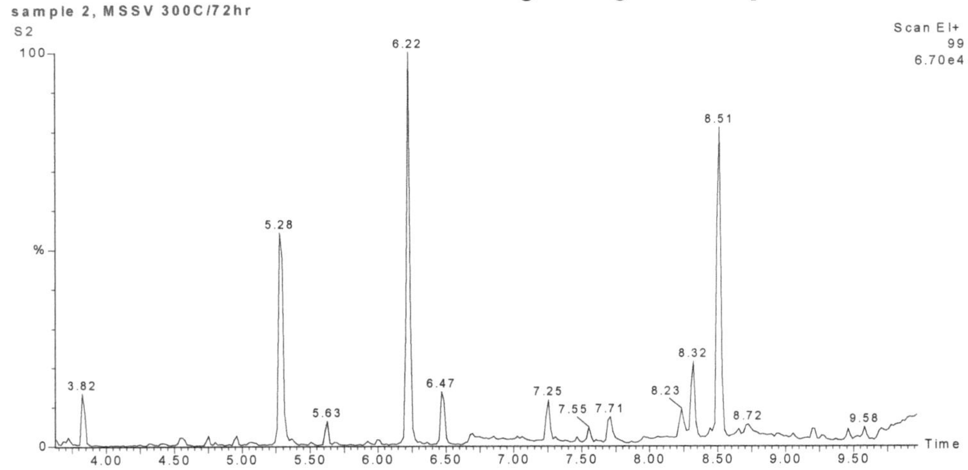

Nubian hair chromatogram diagnostic for lipids
Library Specimen

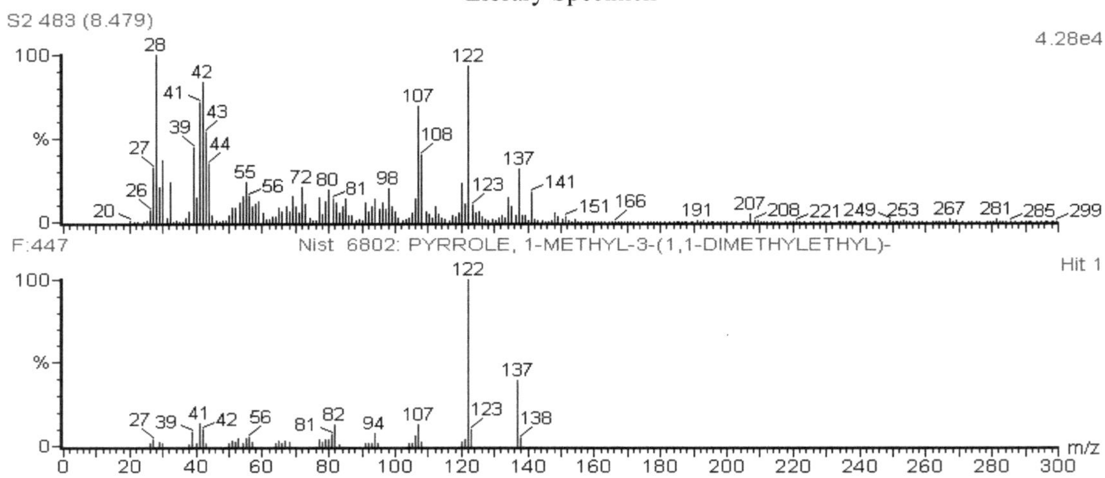

Table 54: ***Reliability (intra-observer variability.)***

VARIABLE	*NO OF PAIRS*	*KAPPA (Standard Error)*	*RELIABILITY COEFFICIENT*	*MEAN DIFFERENCE (95% CI) between 1st and 2nd measurement*
Gender	*10*	*0.737 (0.241)*	-	-
Age group	*9*	*1.00*	-	-
Number of teeth	*10*	-	*1.00*	*0.0*
Antemortem loss	*10*	-	*0.994*	*-0.1 (-0.33; -0.13)*
Wear 1st molar	*10*	*0.873 (0.112) #*	-	-
Wear 2nd molar	*9*	*0.946 (0.059) #*	-	-
Wear 3rd molar	*8*	*1.00*	-	-
Condyle m/l in mm.	*9*	-	*0.943*	*-0.047 (-0.34; 0.24)*
Form + Surface *Temporomandibular joint*	*9*	*0.842 (0.08) #*	-	-
Continuous eruption	*8*	-	*0.779*	*-0.81 (-0.77; 0.61)*
Clinical crown	*8*	-	*0.721*	*0.30 (-0.32; 0.92)*
Caries	*10*	*1.00*	-	-
Horizontal *Periodontal breakdown*	*10*	*1.00*	-	-
Vertical *Periodontal breakdown*	*10*	*1.00*	-	-
Abscess	*10*	*1.00*	-	-
Cyst	*10*	*1.00*	-	-
Anomalies	*10*	*0.615 (0.337)*	-	-

Table 55. ***Reliability (intra-observer variability.)***

MEASUREMENT*with 2 pairs of calipers 7 specimens*	***RELIABILITY COEFFICIENT***
Condyle m/l	*Reliability coefficient 0.998*
Continuous eruption	*Reliability coefficient 0.996*
Clinical crown	*Reliability coefficient 0.992*

Table 56: ***Reliability (inter-observer*** **variability).**

VARIABLE	*KAPPA*
Gender	***Kappa=1*** *(perfect agreement)*
Age group	***Kappa=1***
Ante mortem loss	***Kappa=1***
Wear 1st molar	***Weighted kappa 0.899***
Wear 2nd molar	***Weighted kappa 0.667***
Condyle m/l	***Reliability coefficient 0.999***
Form + Surface	***Kappa=1***
Continuous eruption	***Reliability coefficient 0.985***
Clinical crown	***Reliability coefficient 0.987***
Caries	***Weighted kappa 0.937***
Periodontal breakdown	***Kappa=1***
Abscess	***Weighted kappa*** *0.391*

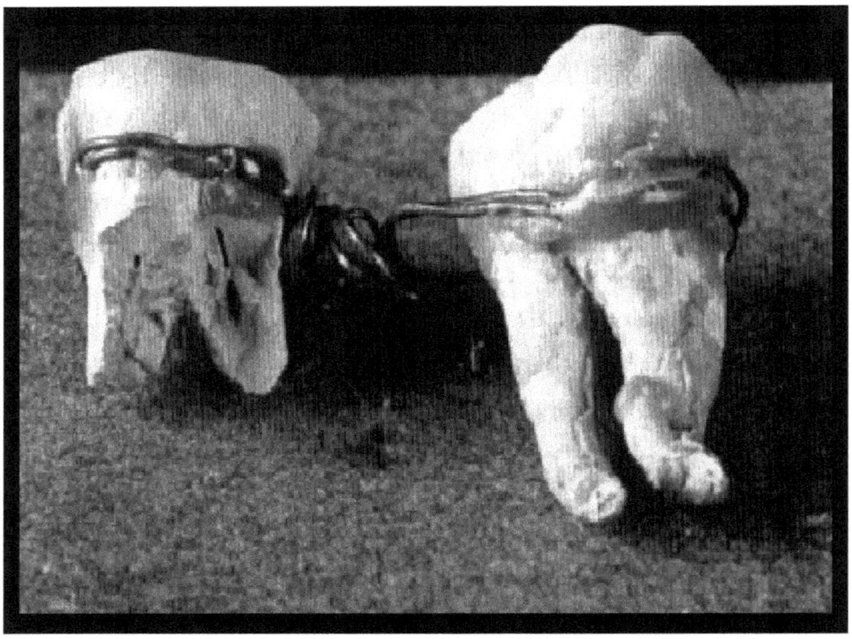

Fig 1: ***Junker's 'bridge'.***
Ref. 3.6.1

Fig. 2a: ***Typical dentition with no evidence of extractions.***
Ref. 3.6.4

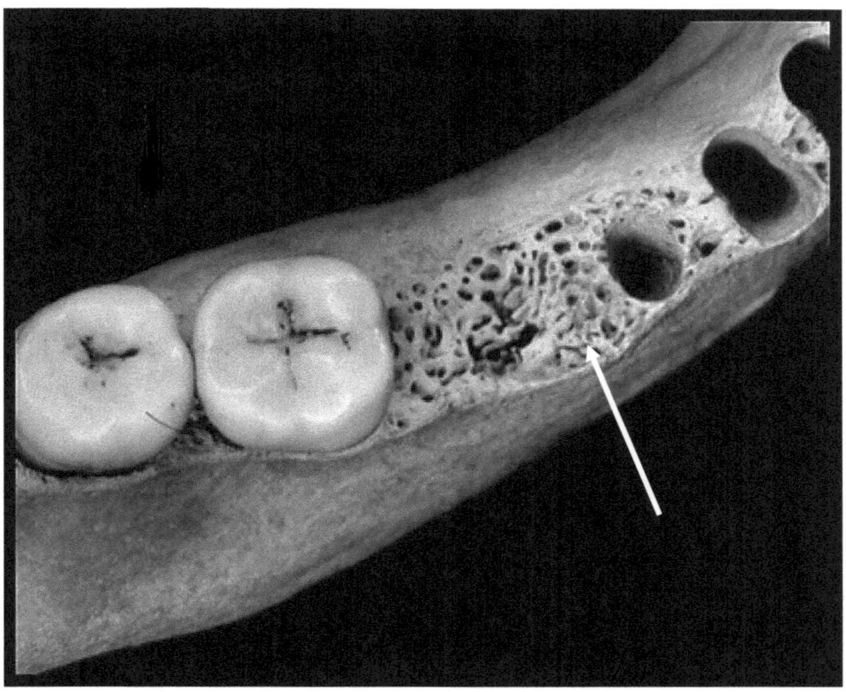

Fig. 2b: ***Dentition with possible evidence of extraction. Healing in 1st molar socket. Also showing postmortem loss of premolars and canine. (Miller, 1997)***
By courtesy of Natural History Museum

Fig 3: ***Mandible& skull showing male characteristics***
Ref. 4.2.1

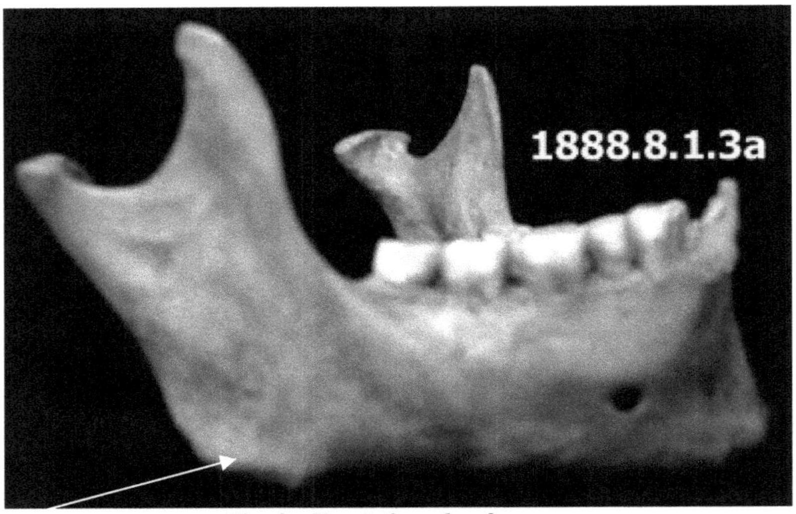

Fig. 3a ***Everted angle of ramus.***

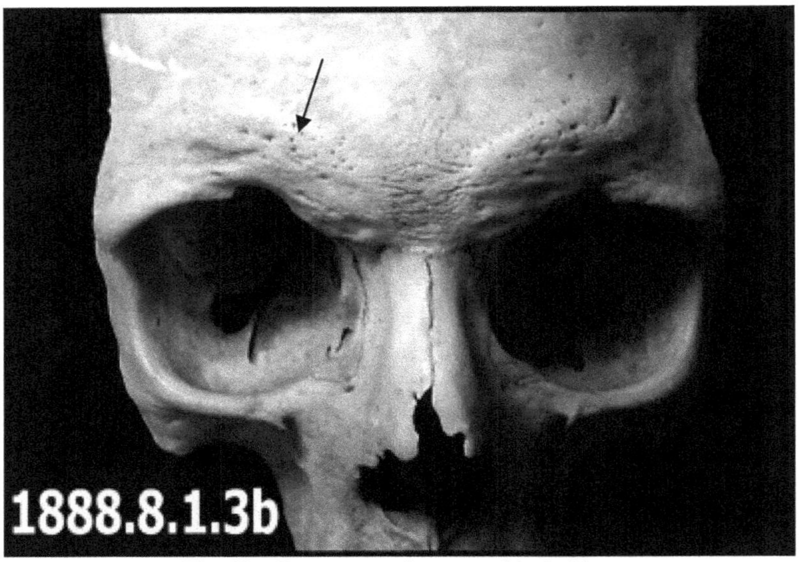

Fig. 3b: ***Pronounced supraorbital ridge.***

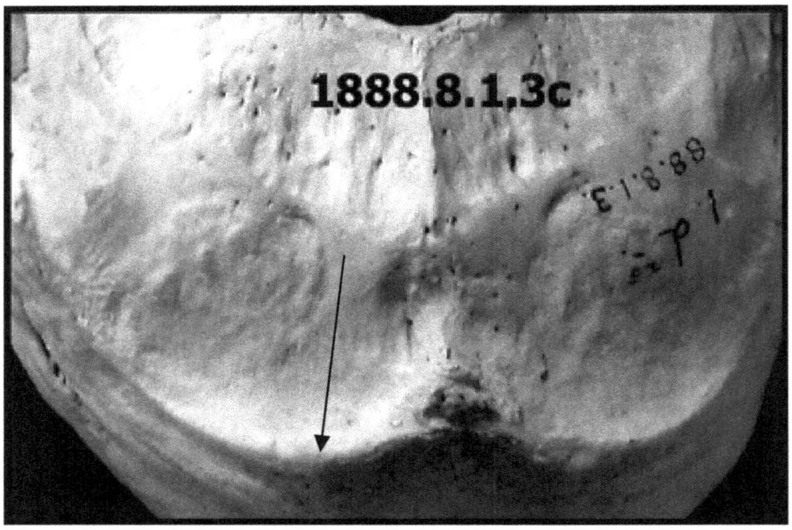

Fig. 3c: ***Pronounced nuchal crest***

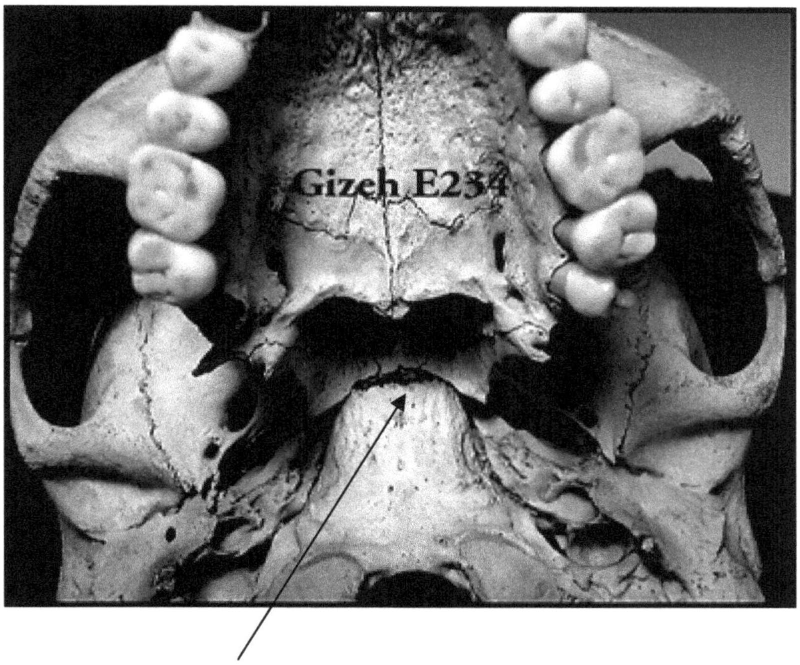

Fig 4a: ***Open spheno-occipital synchondrosis with mild wear of occlusal surfaces.***
Ref. 4.3.1

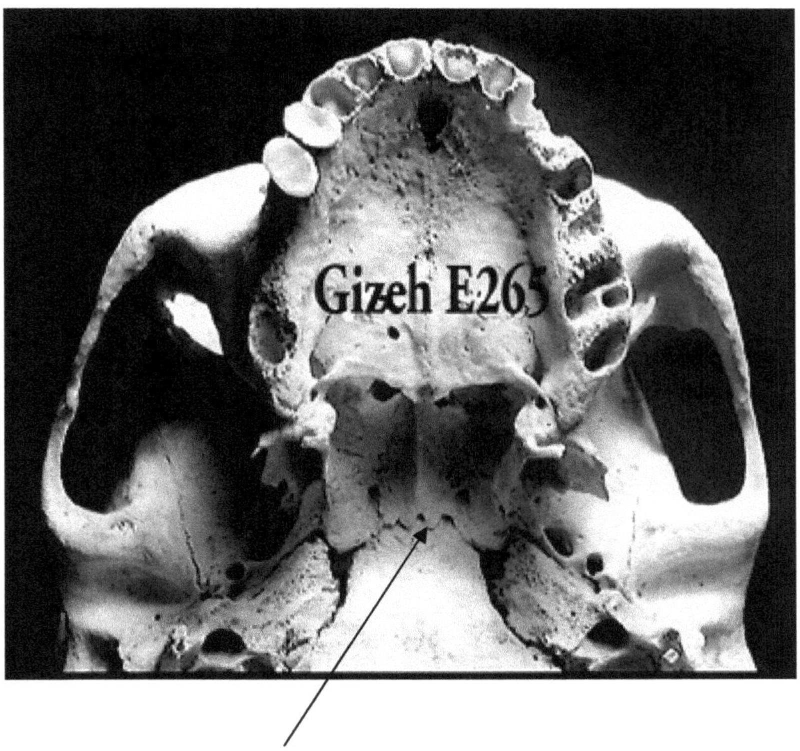

Fig 4b: ***Closed spheno-occipital synchondrosis. Ante mortem tooth loss and occlusal wear.***

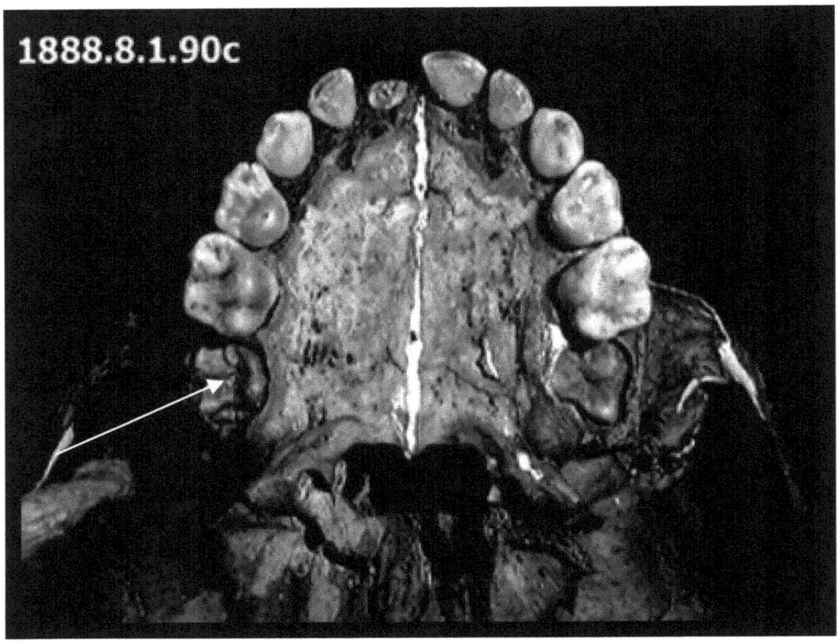

Fig 5a: ***Dentition shows development of 6/6 in crypts. All primary dentition present.***

Ref. 4.3.2.1

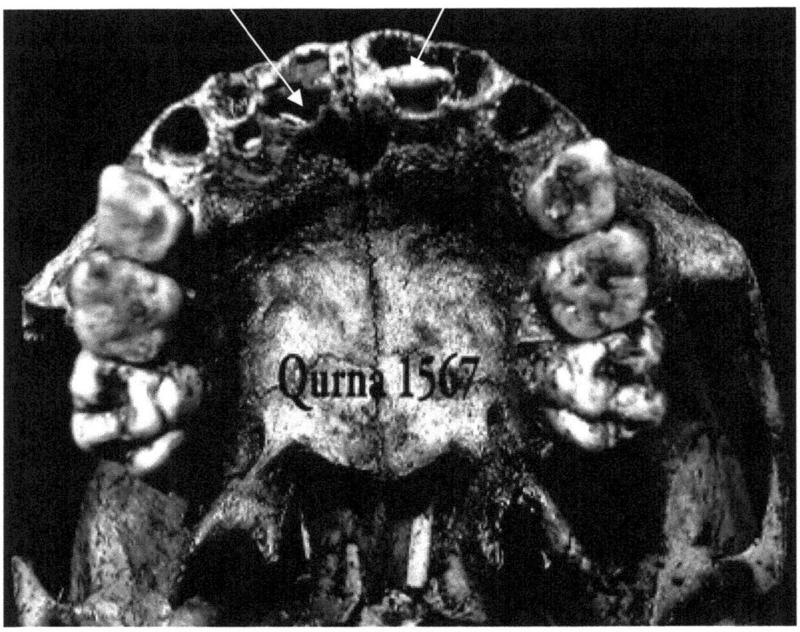

Fig 5b: ***6/6 in situ and 1/1 erupting.***

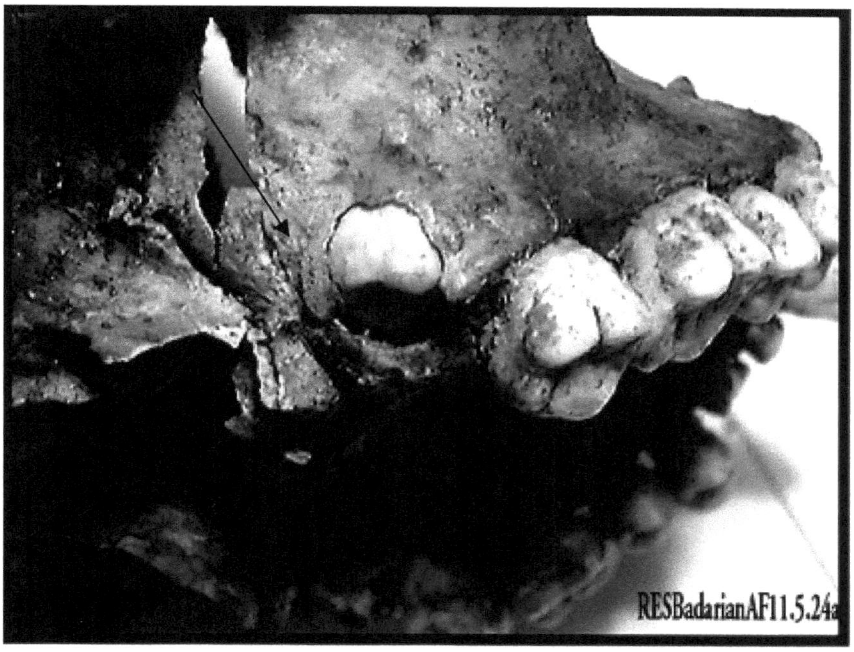

Fig 5c: ***Developing 8/.***
Ref. 4.3.2.1

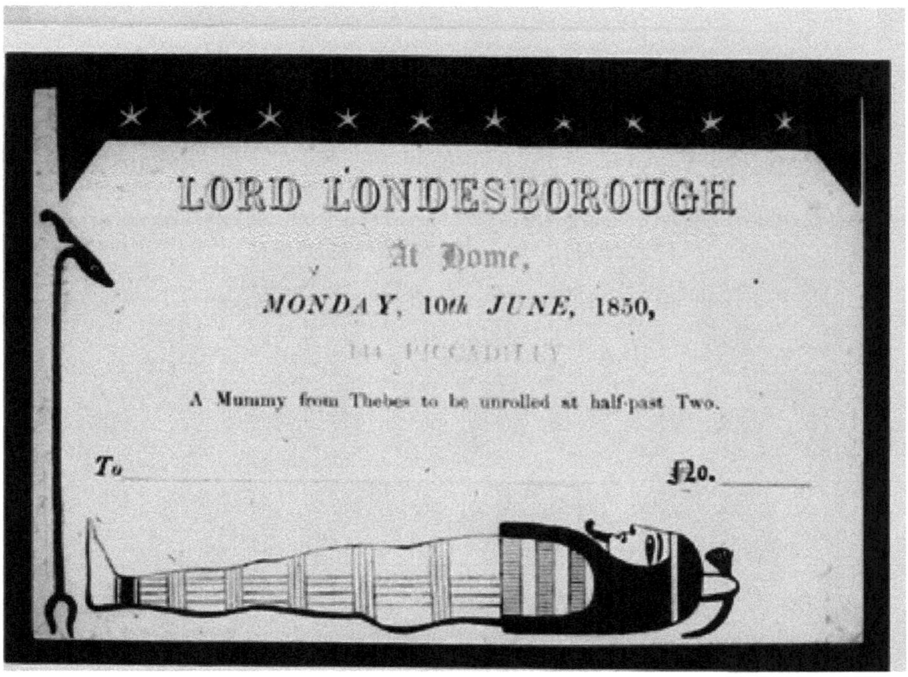

LORD LONDESBOROUGH

At Home,

MONDAY, 10*th* *JUNE*, 1850,

A Mummy from Thebes to be unrolled at half-past Two.

To No.

Fig. 6: ***An invitation to Lord Londesborough's 'At Home'.***

By courtesy of Professor G.T.Martin.

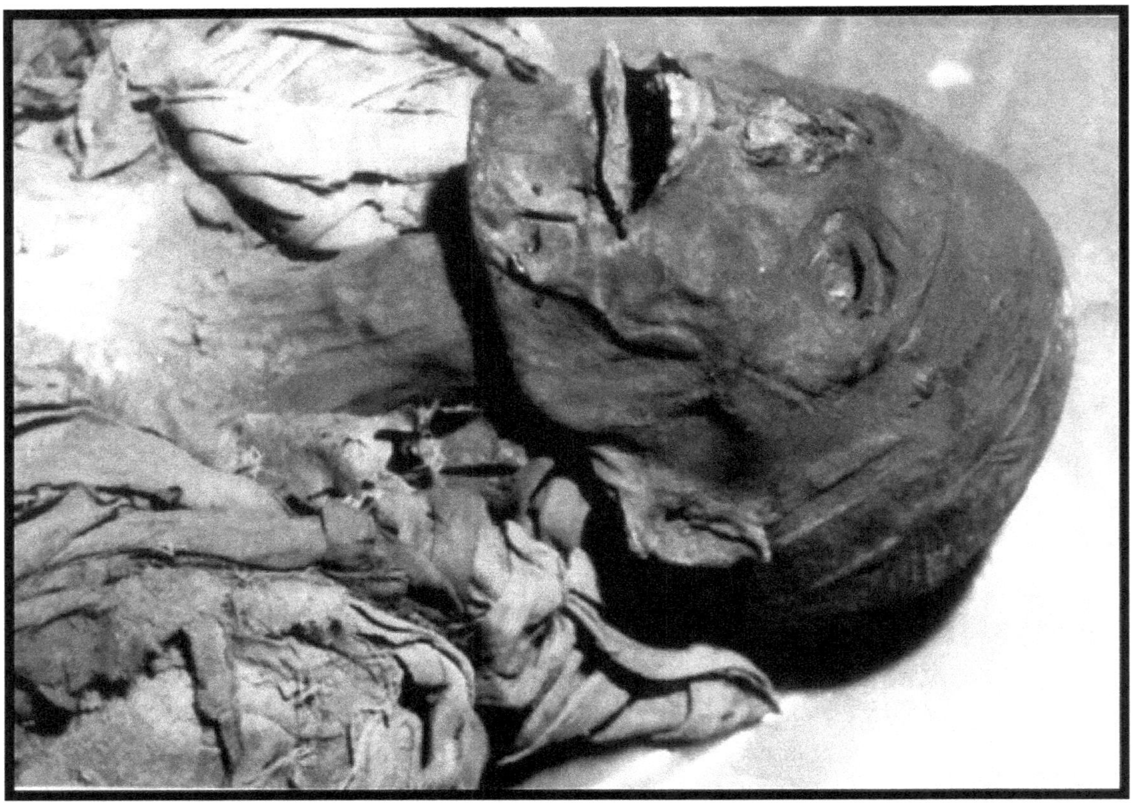

Fig. 7a: ***Natsef Amun with open mouth and projecting tongue.***
Ref. 4.10

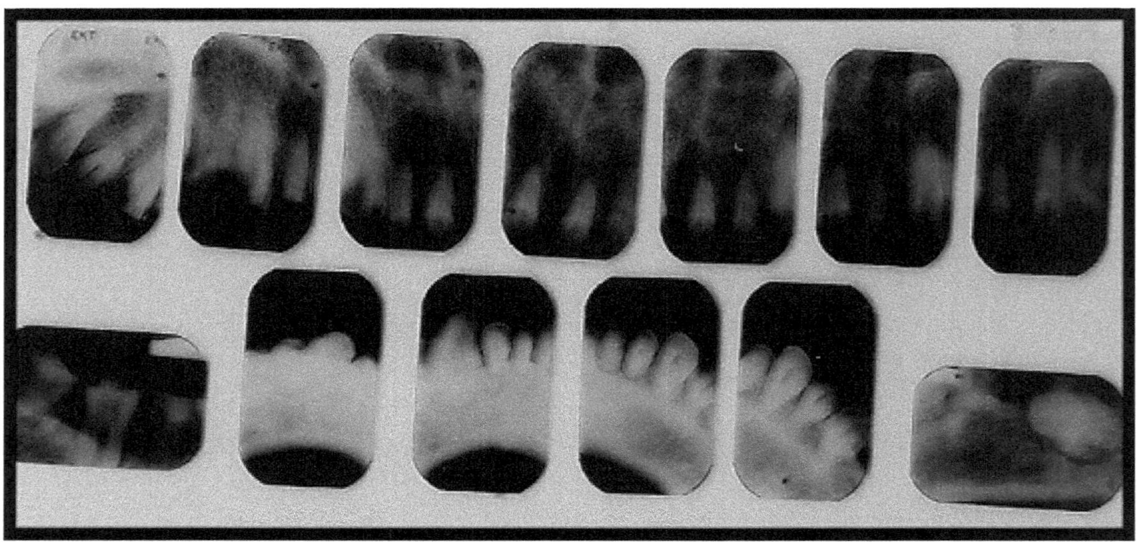

Fig. 7b: ***Intraoral radiographs of Natsef Amun.***

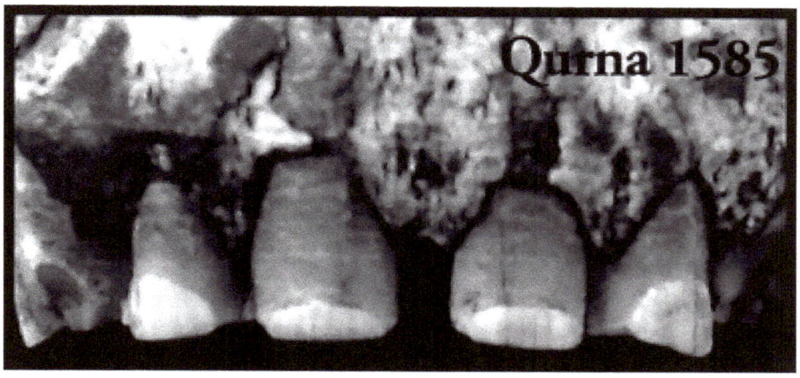

Fig.8a & 8b: ***Tooth surface loss, suggestive of erosion – labial & occlusal views.***
Ref. 4.11.3

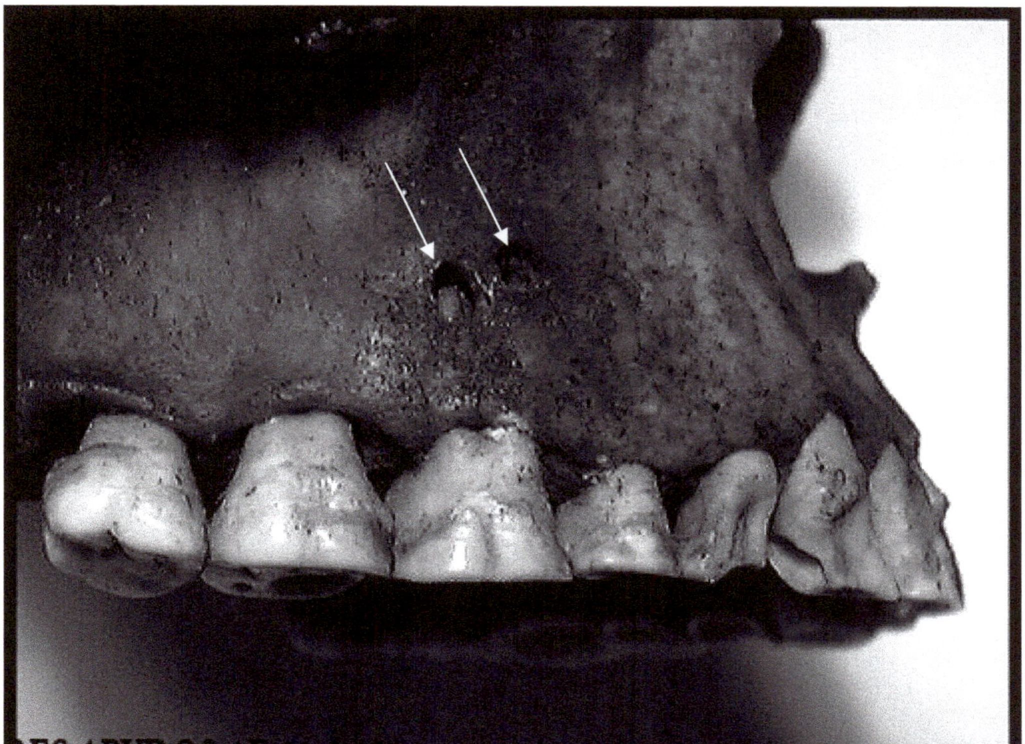

Fig 9: ***Example of periapical lesions which has been interpreted as drainage of abscess by 'fire drill'.***
Ref. 3.6.5

Fig 10: ***Map of Egypt showing sites of specimens*** (not to scale).
Ref. 5.3

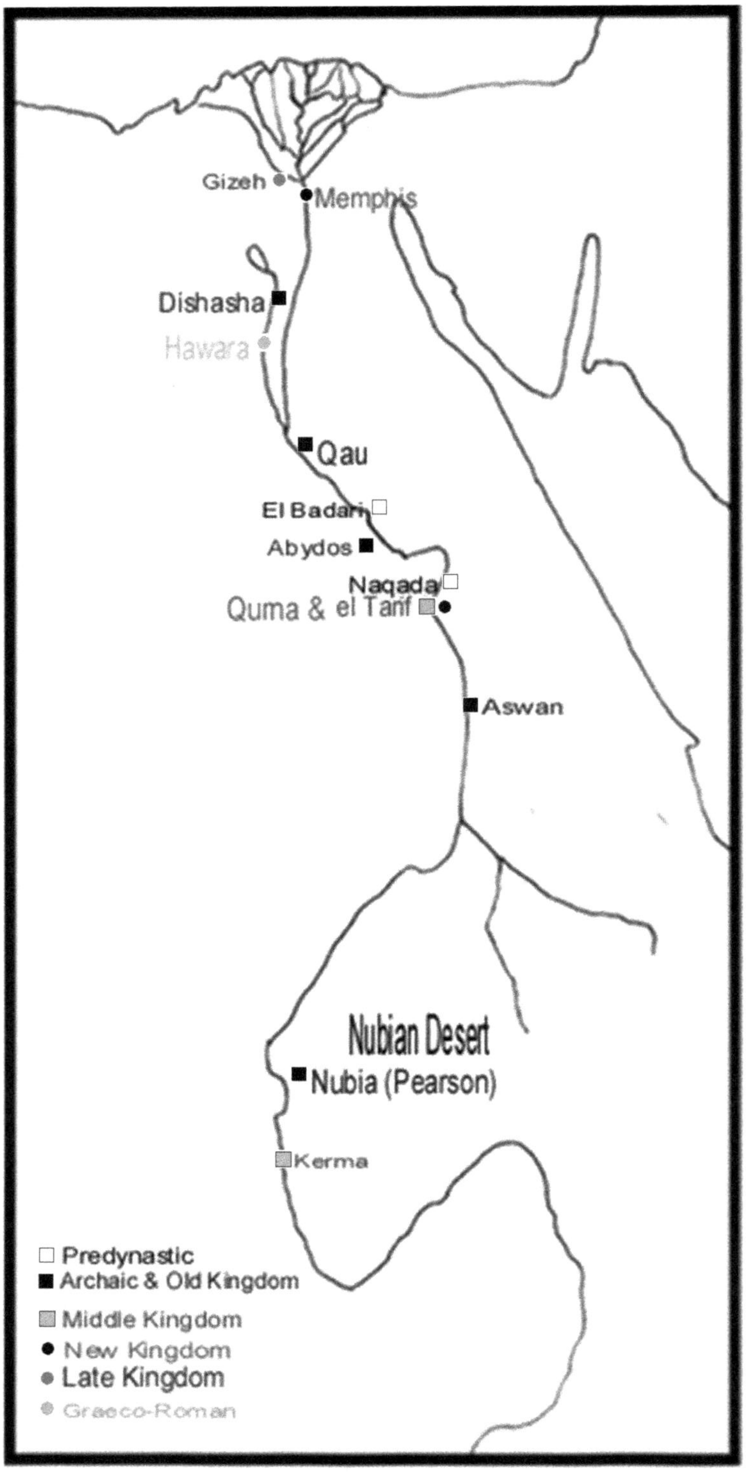

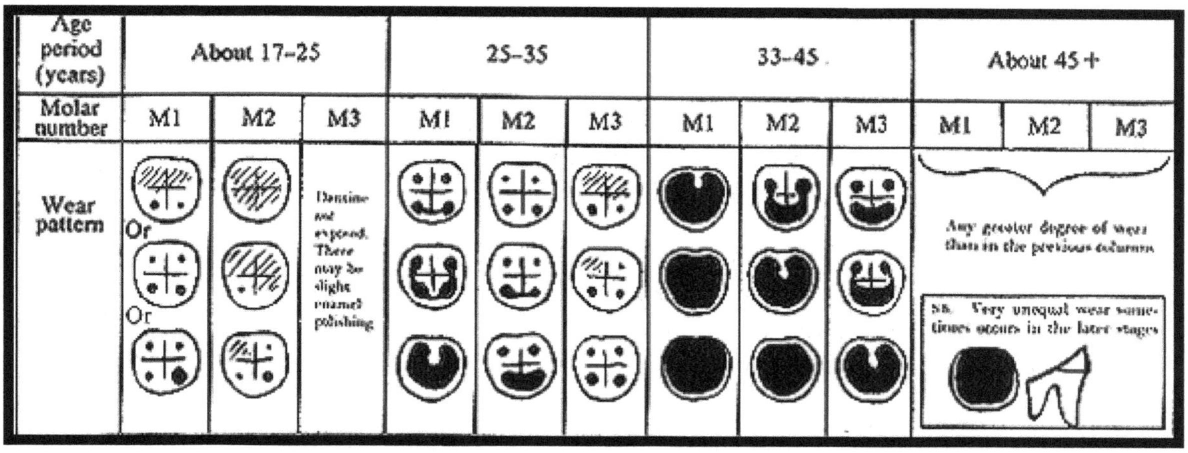

Fig 11: ***Table to assess age by wear.***

Ref. 5.5.1

(By kind permission of Professor D. Brothwell)

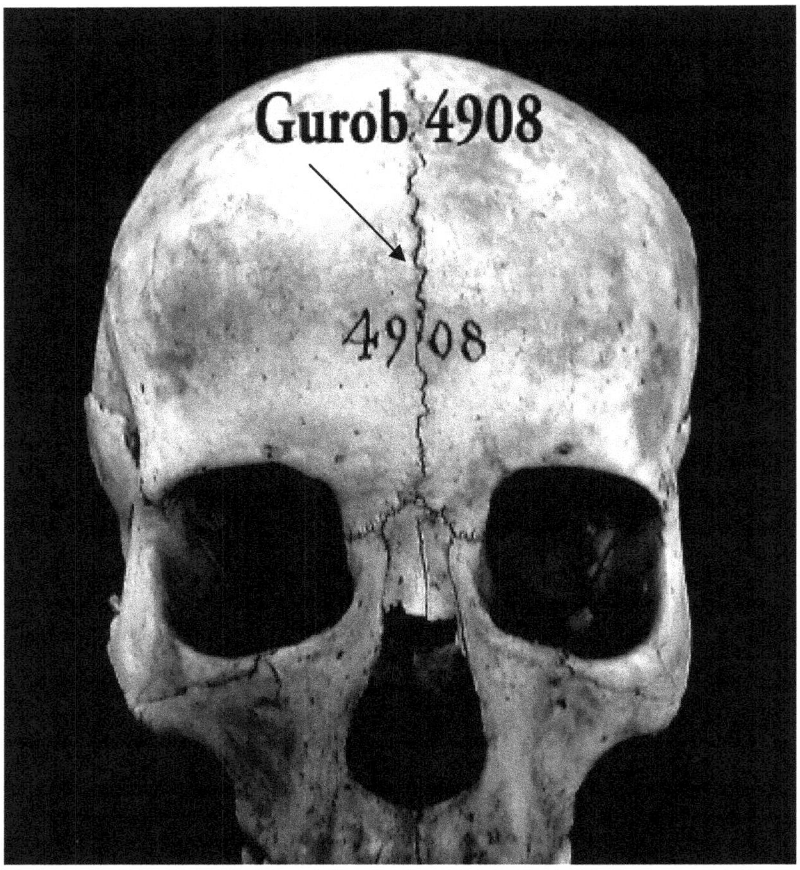

Fig 12: ***Metopic Suture.***

Ref. 7.2.1

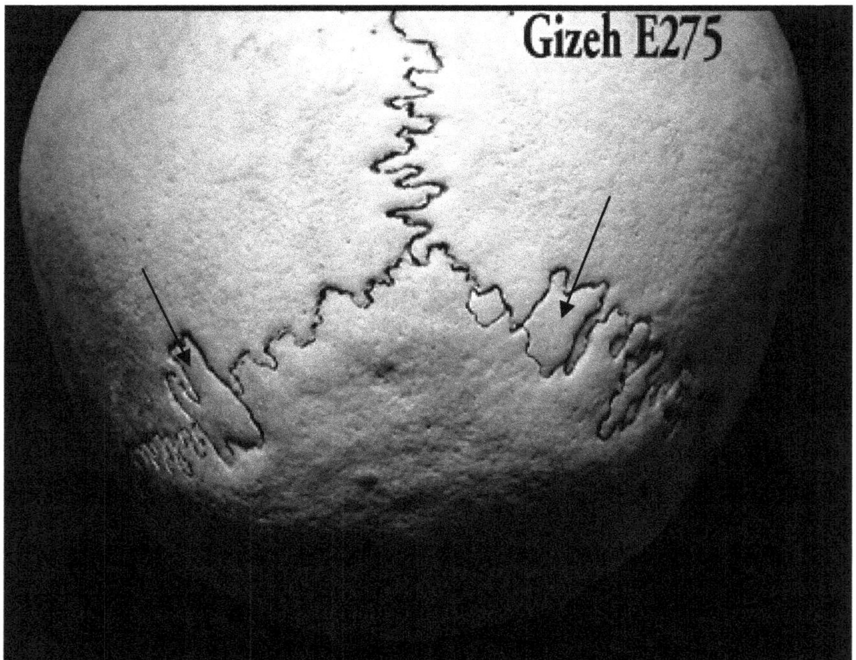

Fig 13: ***Wormian bones.***

Ref. 7.2.2

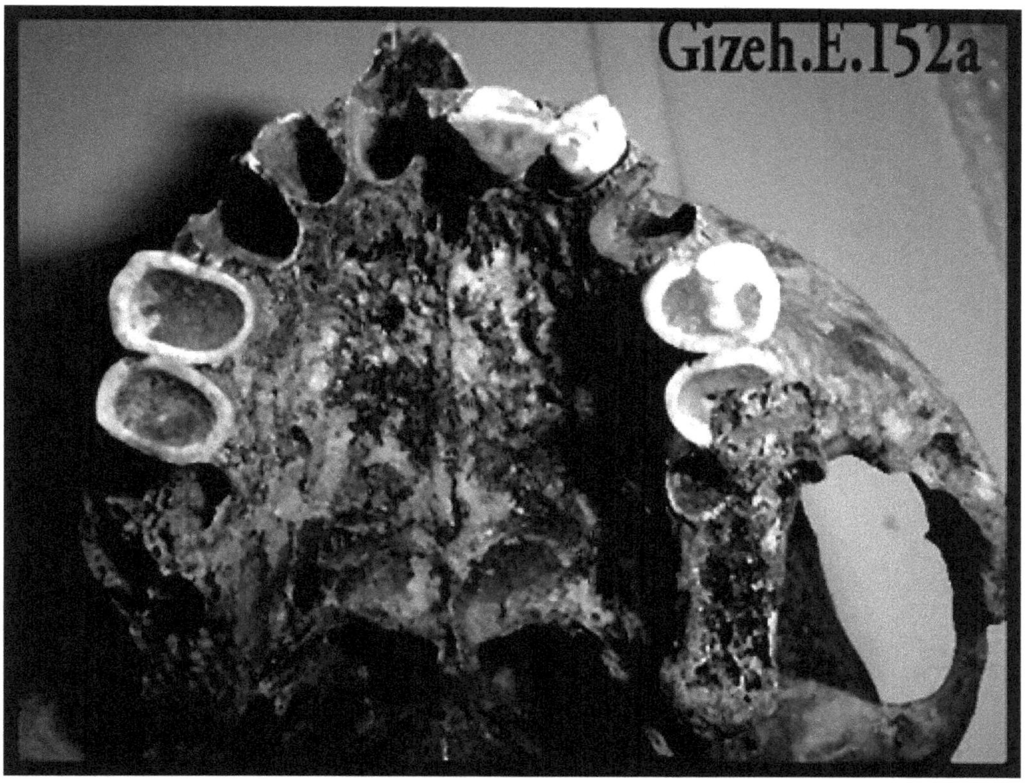

Fig 14: ***Mismatched maxilla (severe wear)& mandible (no wear) from Duckworth Collection.***

Ref. 7.3.1.4

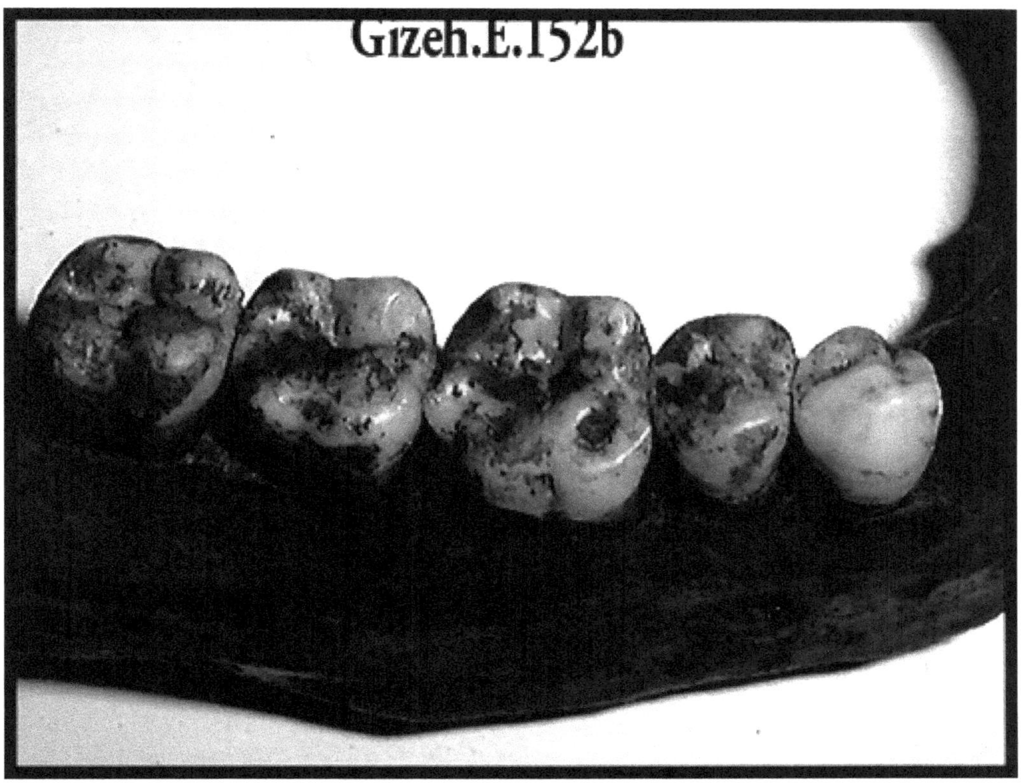

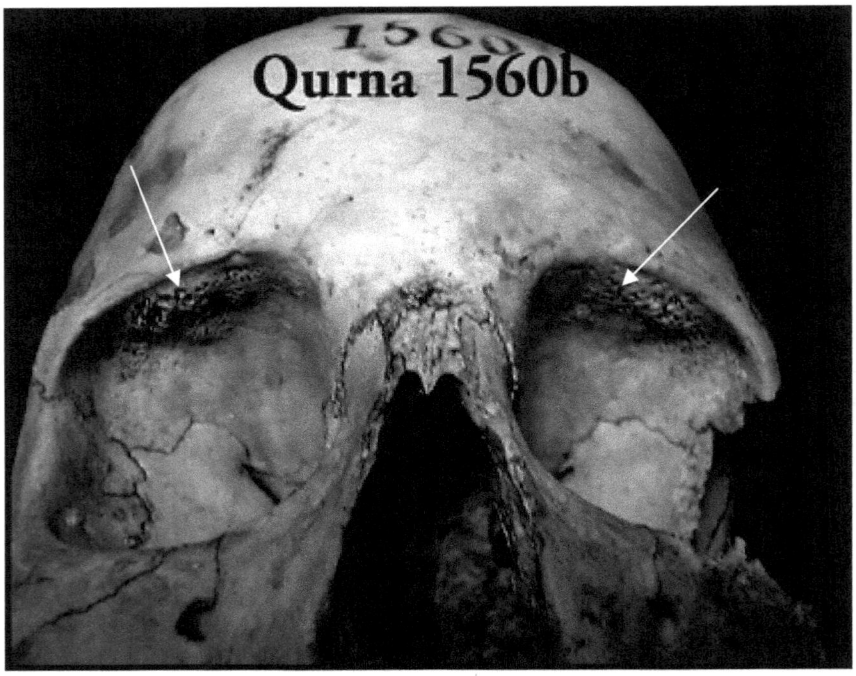

Fig. 15: ***Cribra orbitalia.***

Fig. 7.3.2

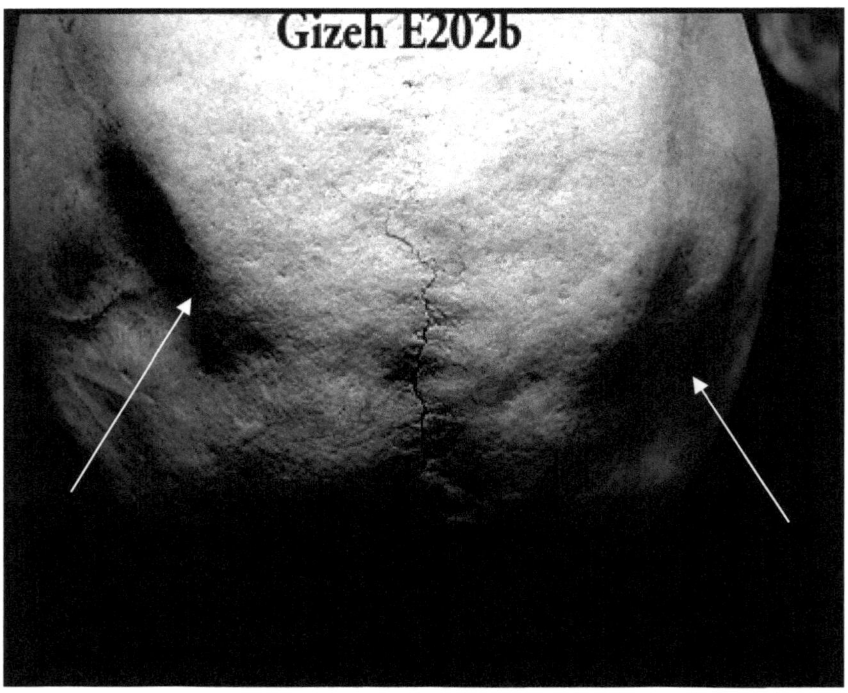

Fig. 16: ***Biparietal depressions.***

Fig. 7.3.2.1

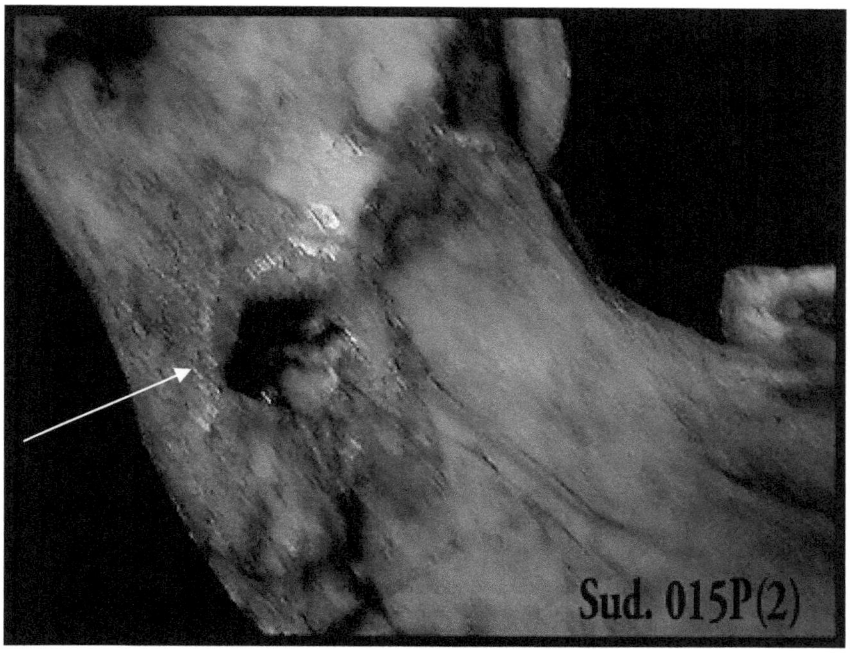

Fig.17a: ***Lesion suggestive of neurofibroma involving inferior dental nerve.***

Ref. 7.3.2.4

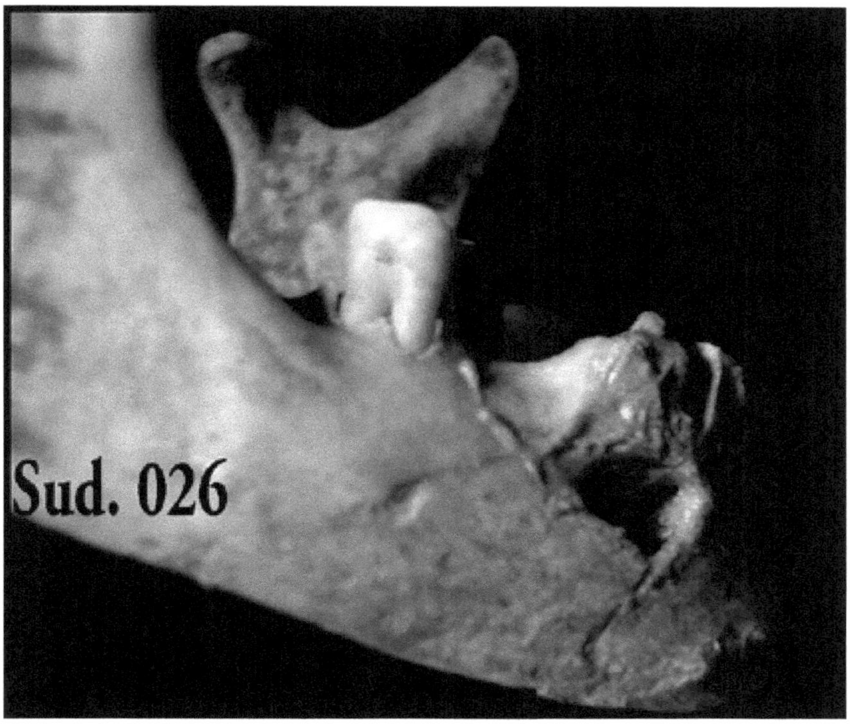

Fig.17b: ***Lesion suggestive of neurofibroma involving mental nerve.***

(More common in mandible than maxilla and is associated with the mental or inferior dental nerve).

Ref. 7.3.2.4

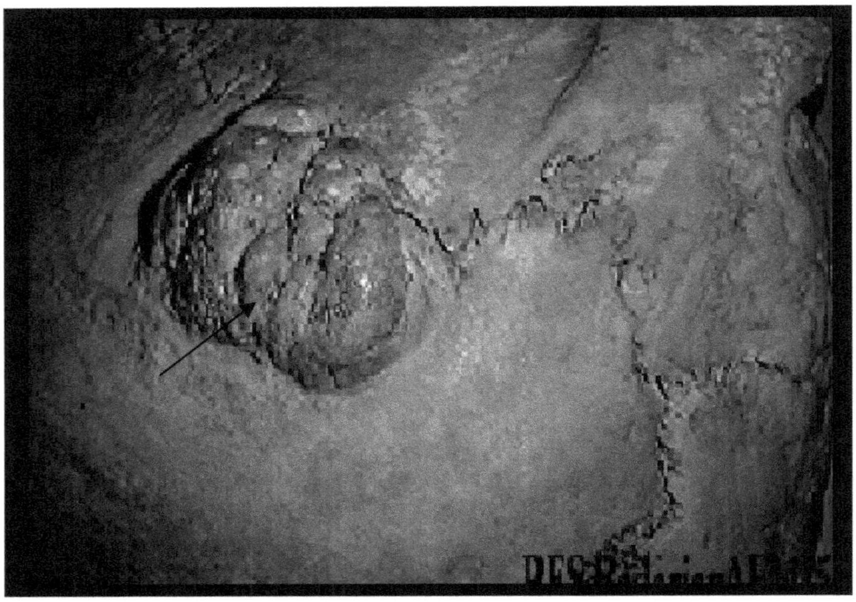

Fig. 18: ***Osteoma.***

Ref. 7.3.2.3

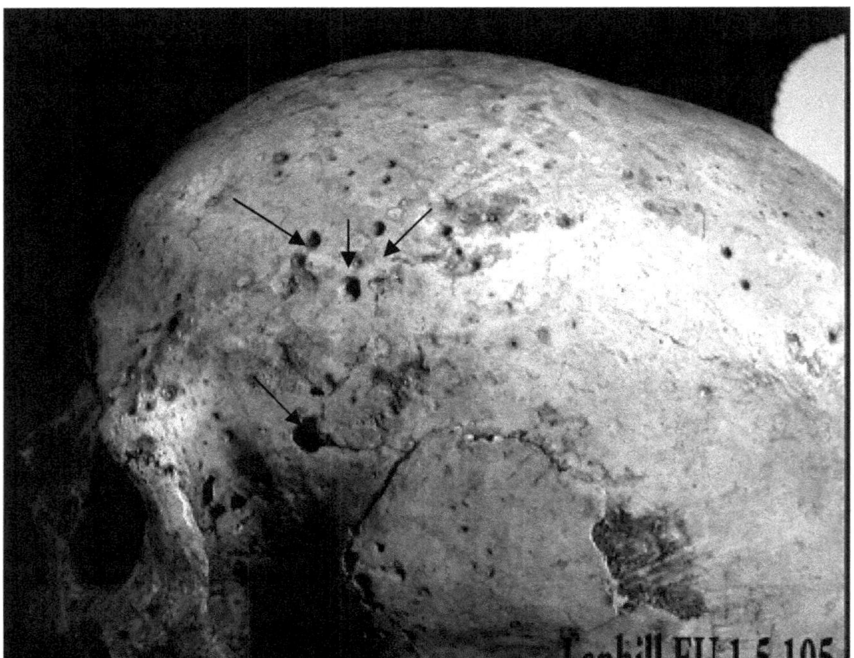

Fig. 19: ***Skull lesions indicative of multiple myeloma, postmortem insect damage, metastatic tumours or hyperparathyroidism?***

Ref. 7.3.1.3 & 7.3.2.5

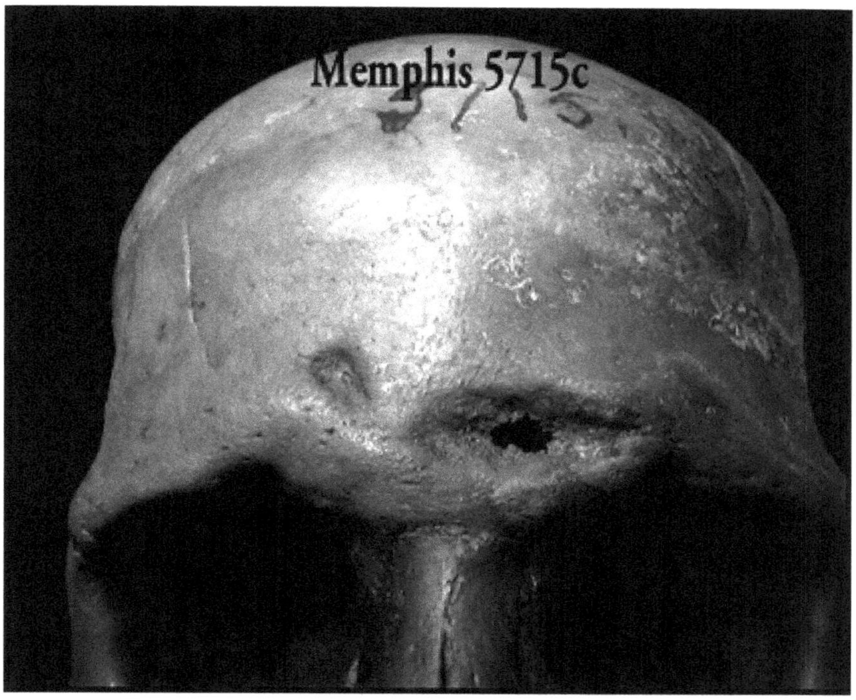

Fig. 20a: ***Healing wound on frontal bone.***

Ref. 7.3.3 & 7.4.4

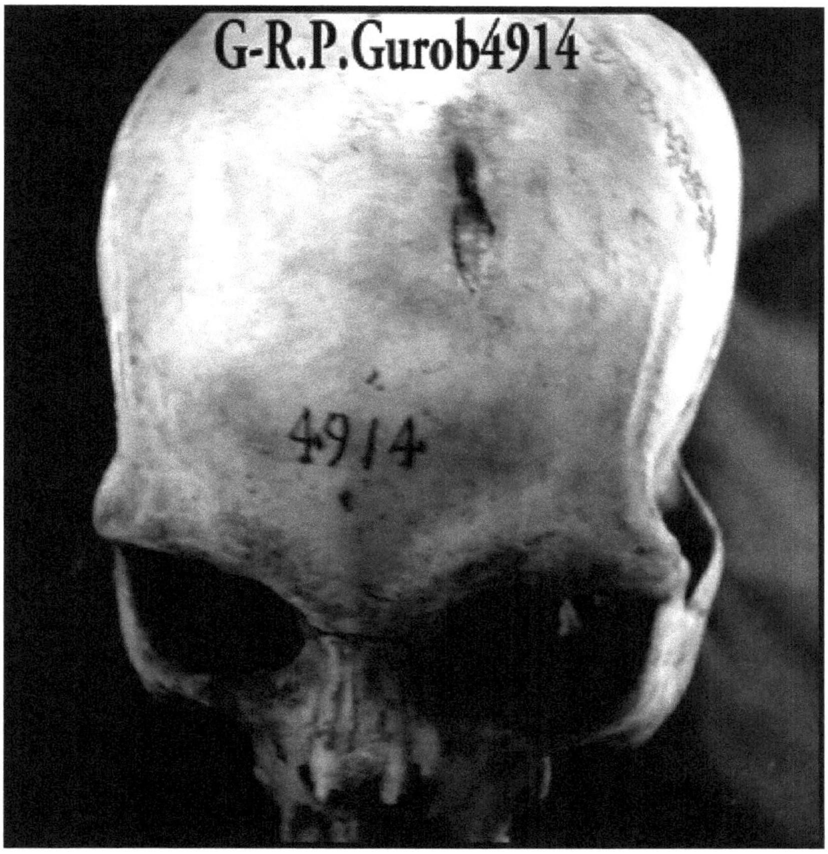

Fig. 20b: ***Wound.***

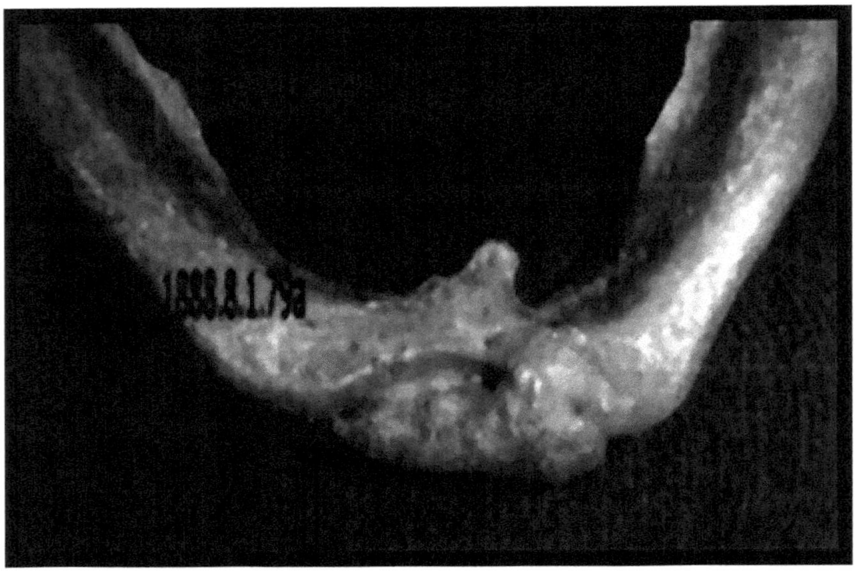

Fig. 20c: ***Healed injury on under surface of mandible.***

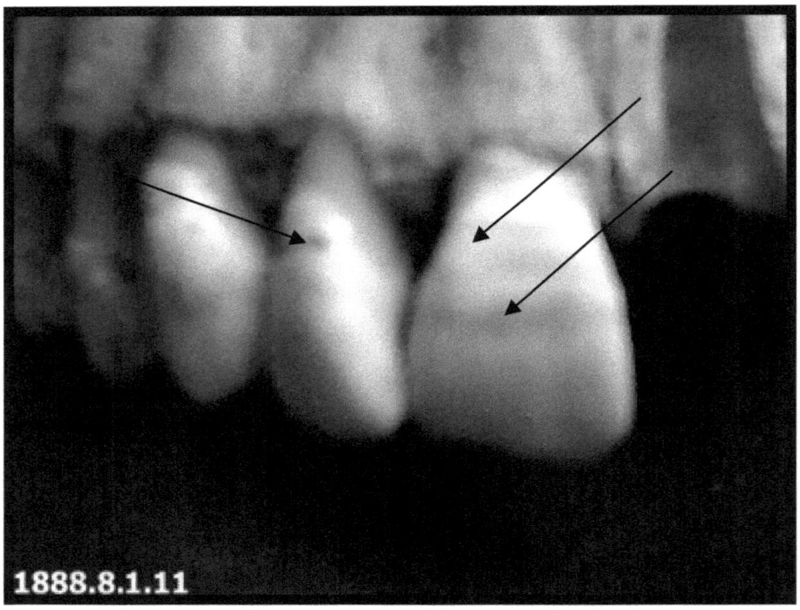

Fig. 21: ***Hypoplasia.***

Ref. 7.4.1

Fig. 22: *Hawara portrait EE.*
There had been difficulty identifying gender of the skull for which this was alleged to be the portrait.

Ref. 7.5.1

(By kind permission of Dr. Paul Roberts from the British Museum.)

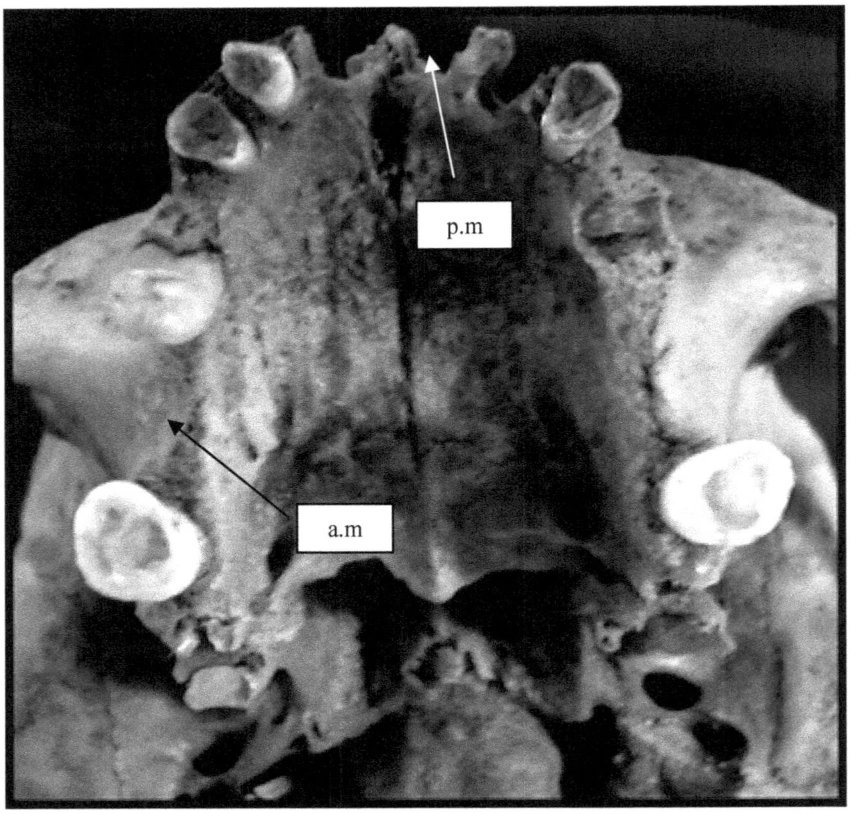

Fig. 23a: ***Ante mortem (a.m) + post mortem (p.m) tooth loss.***

Ref. 7.6

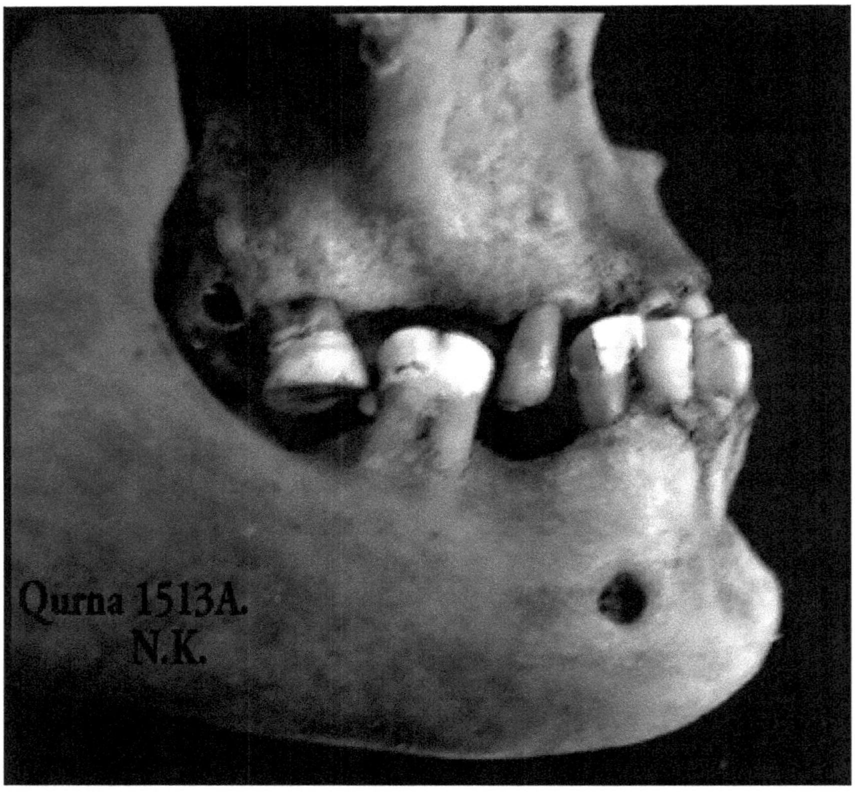

Fig. 23b: ***Over eruption caused by antemortem loss of teeth.***

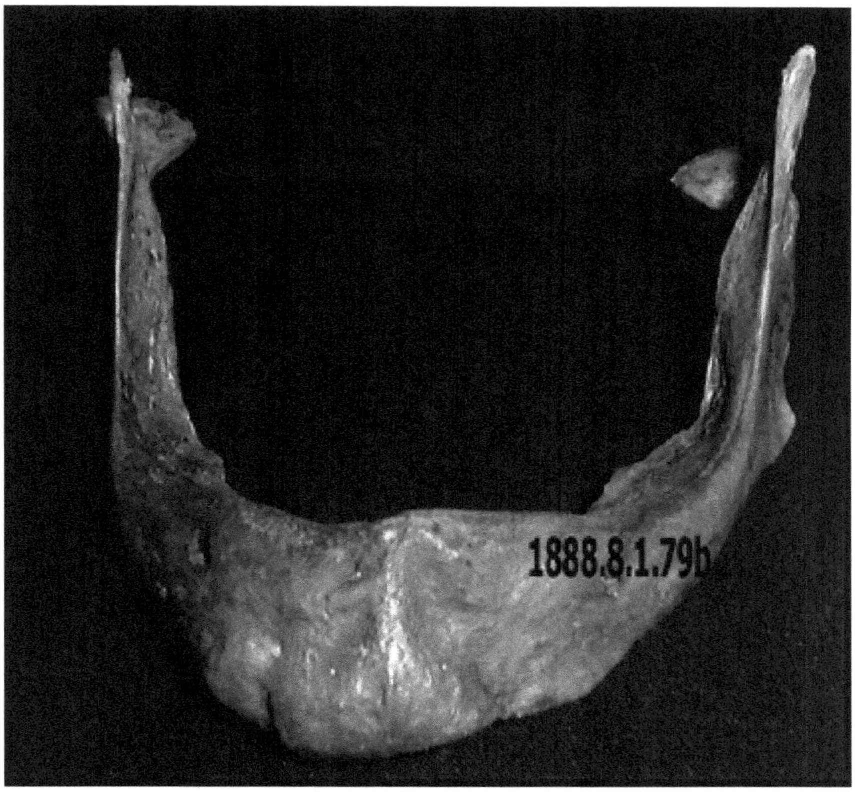

Fig. 23c: ***Ante mortem tooth loss.***

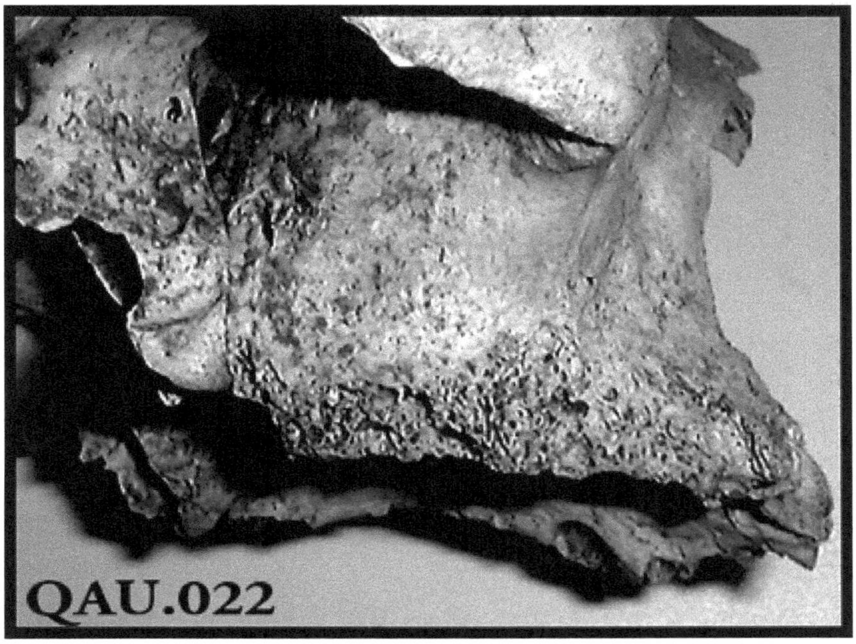

Fig. 23d: ***Ante mortem tooth loss with alveolar damage indicative of osteitis.***

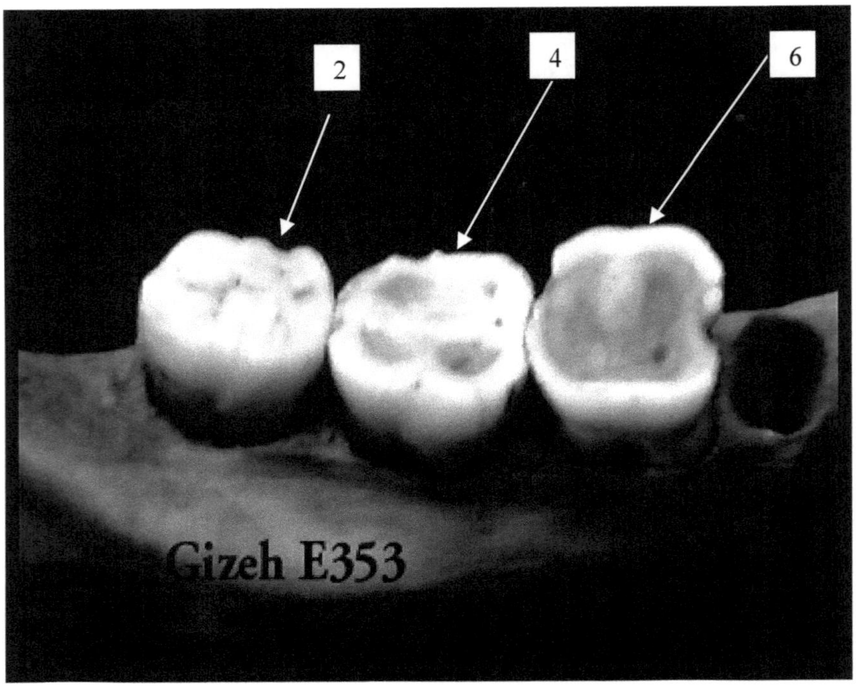

Fig. 24a: ***Example of wear – Molnar grade 6/4/2.***

Ref. 7.7

Fig. 24b: ***Example of wear – Molnar grade 7/5/3.***

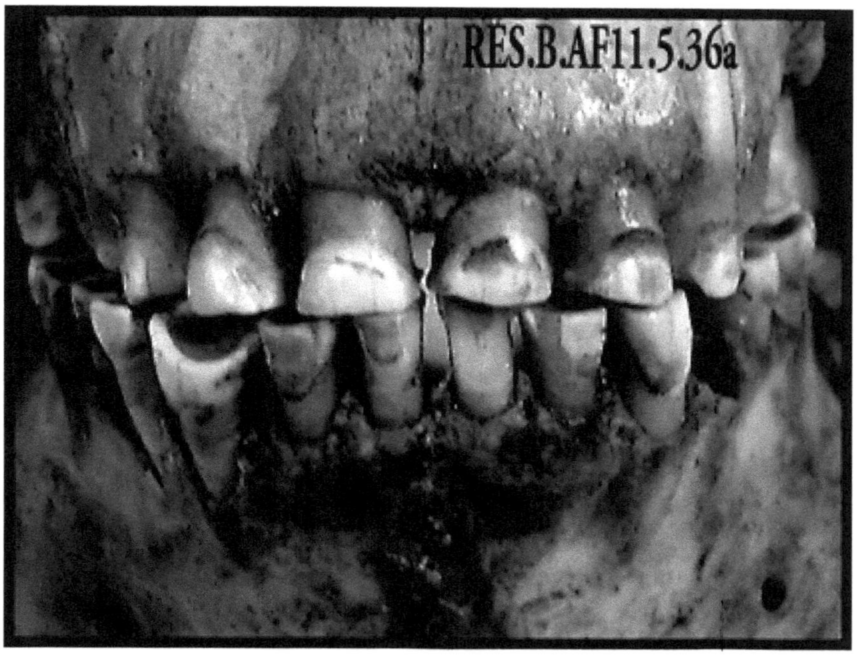

Figs. 24c & d: ***Example of edge to edge anterior occlusion causing severe wear.***

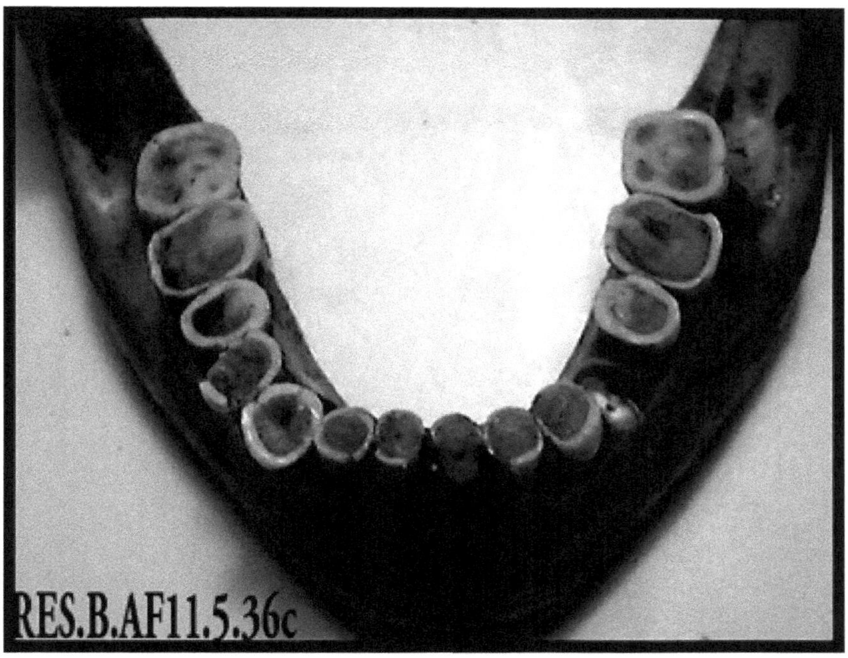

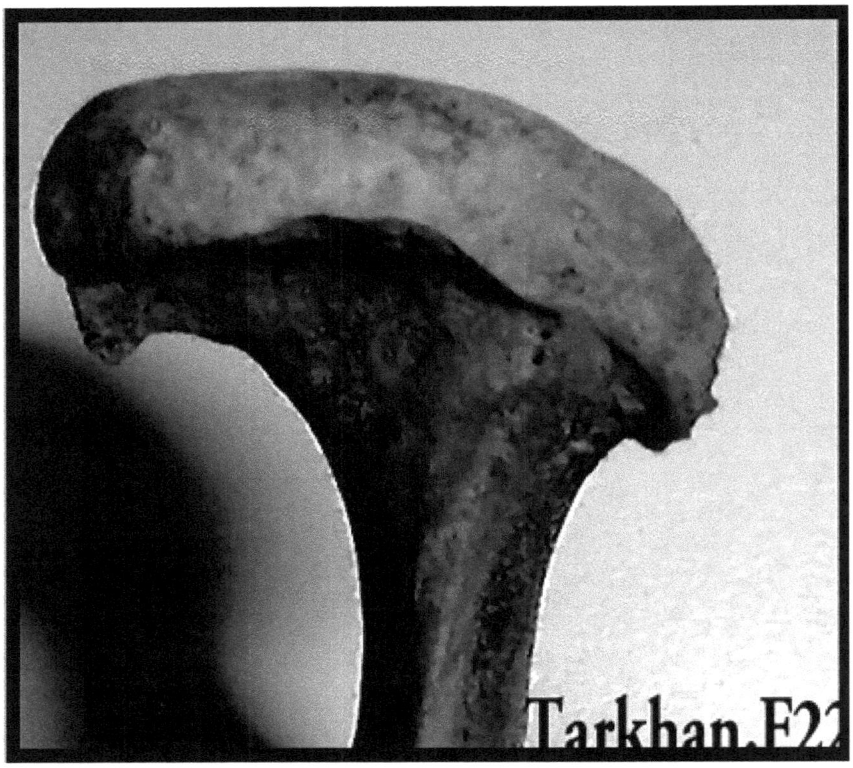

Fig. 25a: ***Osteophyte formation under condyle of temporomandibular joint.***

Ref. 7.9

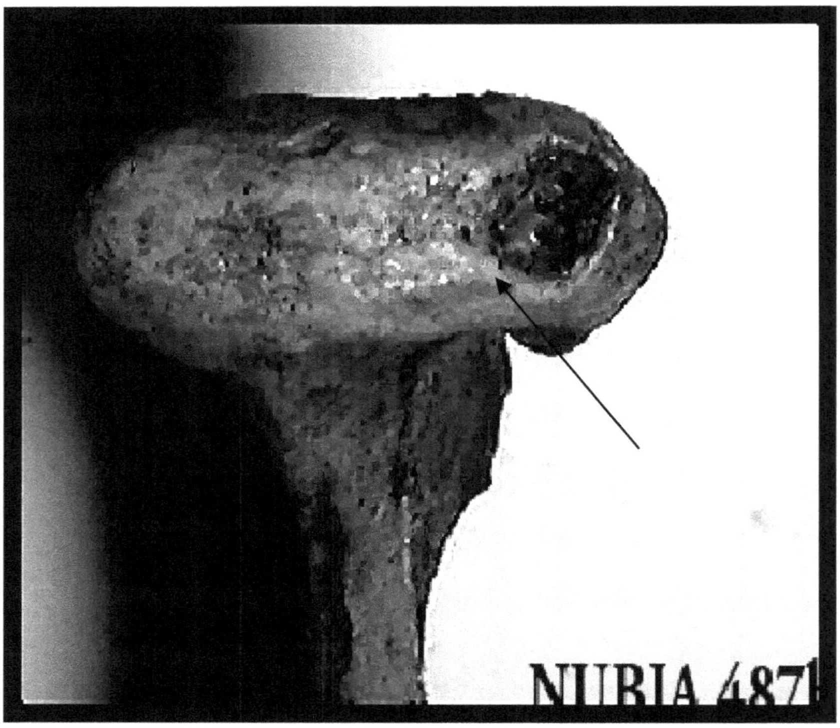

Fig. 25b: ***Erosion on condyle.***

Fig. 25c: ***Form change – deforming change=3***
Surface change – compact layer broken up > 3mm^2 =3
F+ S =6 (Wedel et al.1978).

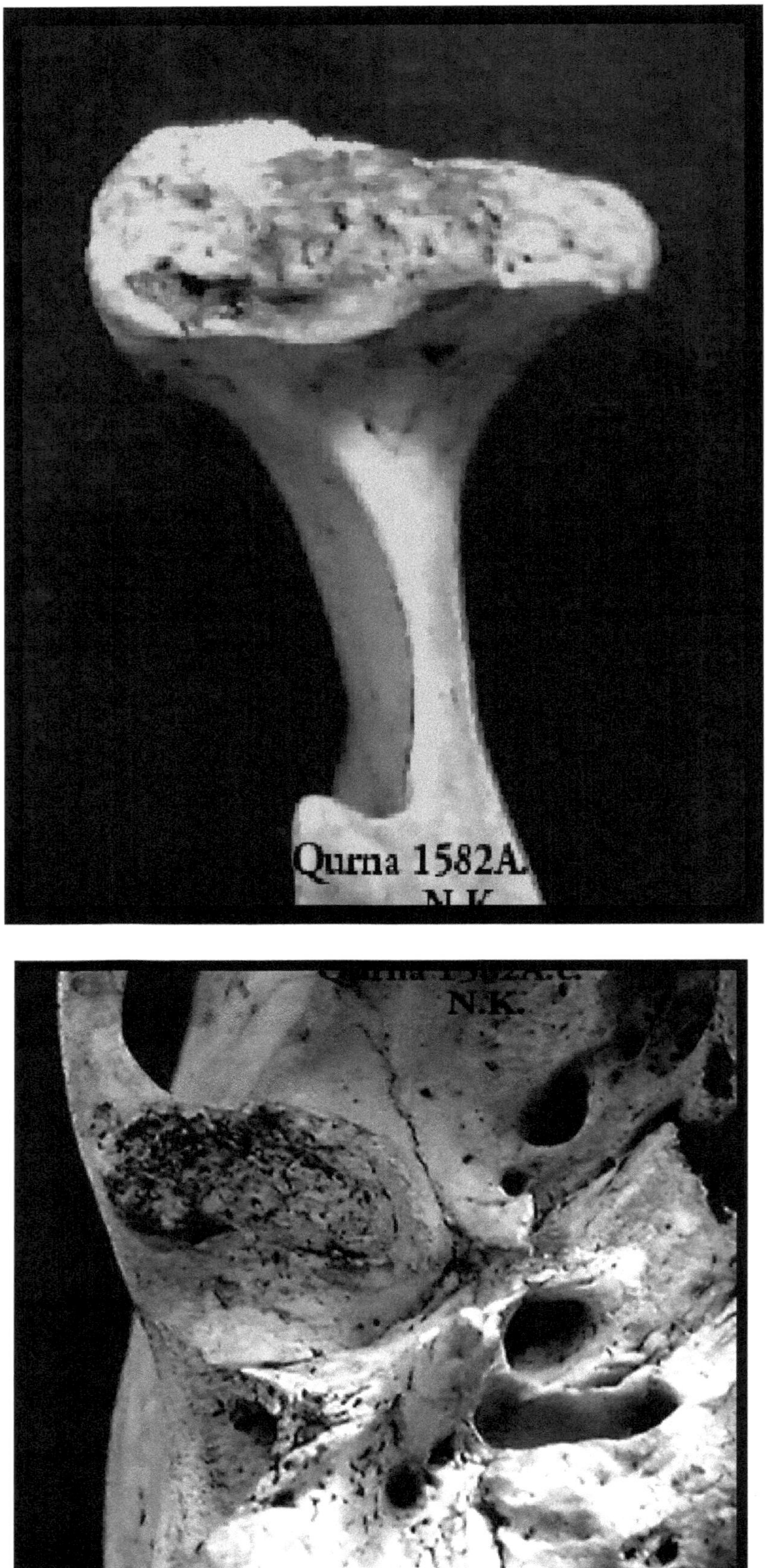

Fig. 25 d & e: ***Condyle & glenoid fossa with signs of severe osteoarthritic change.***

Figs 26: ***Teeth exhibiting cavities arising from caries.***

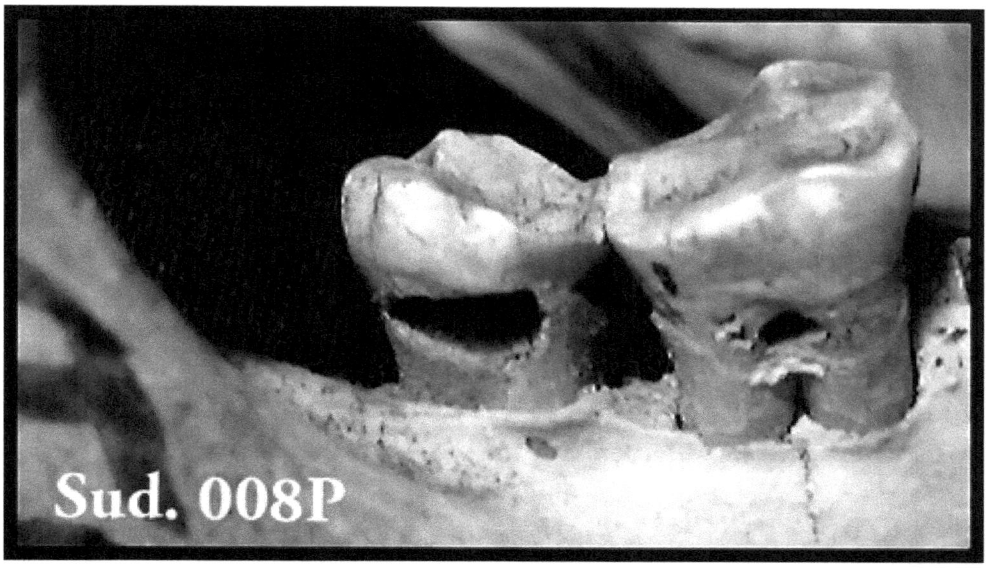

Fig 26a: ***Root caries.***

Ref. 7.10

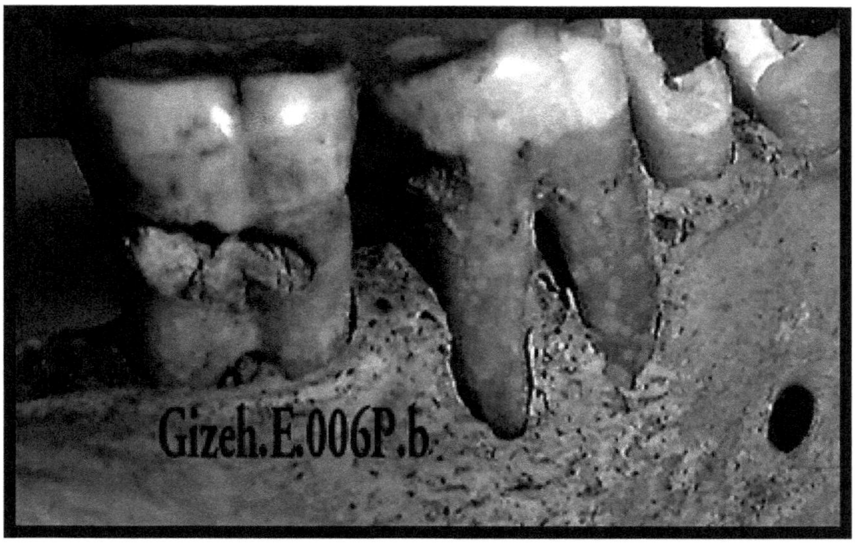

Fig 26b: ***Root caries.***

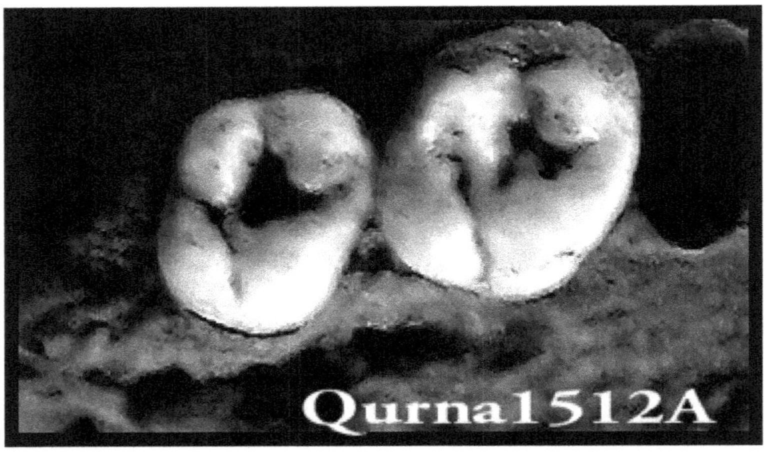

Fig 26c: ***Occlusal cavities.***

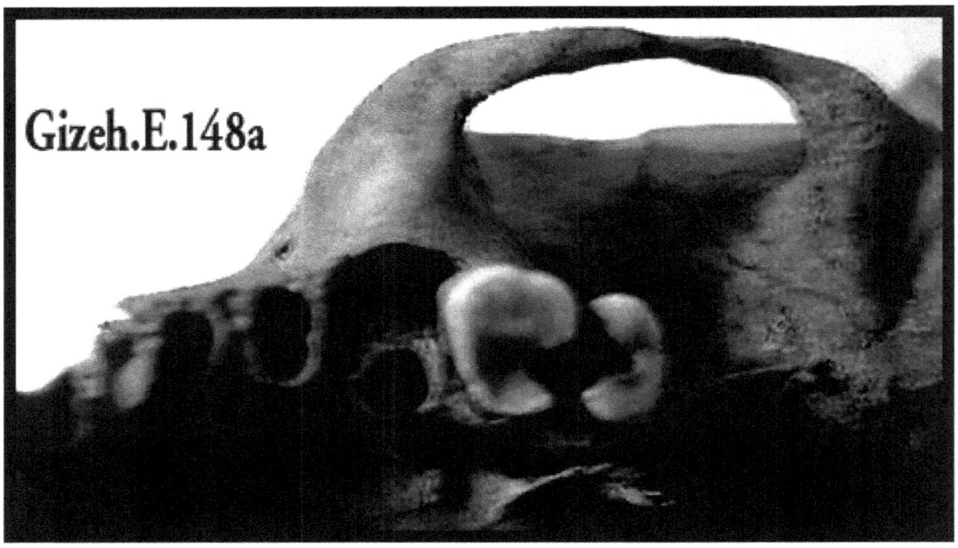

Fig 26 d & e: ***Interstitial cavities.***

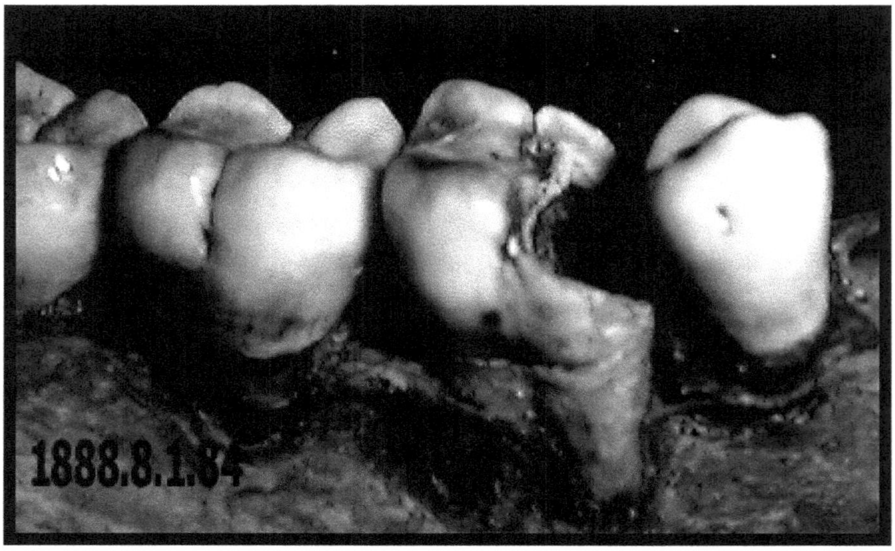

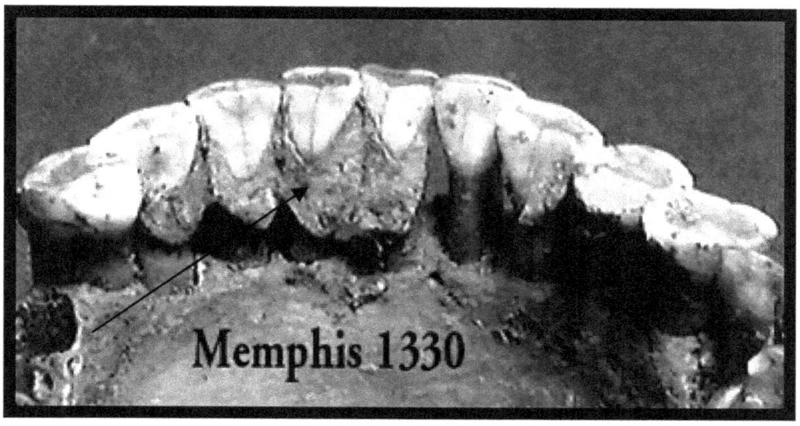

Fig 27a: ***Gross calculus deposits***

Ref. 7.11

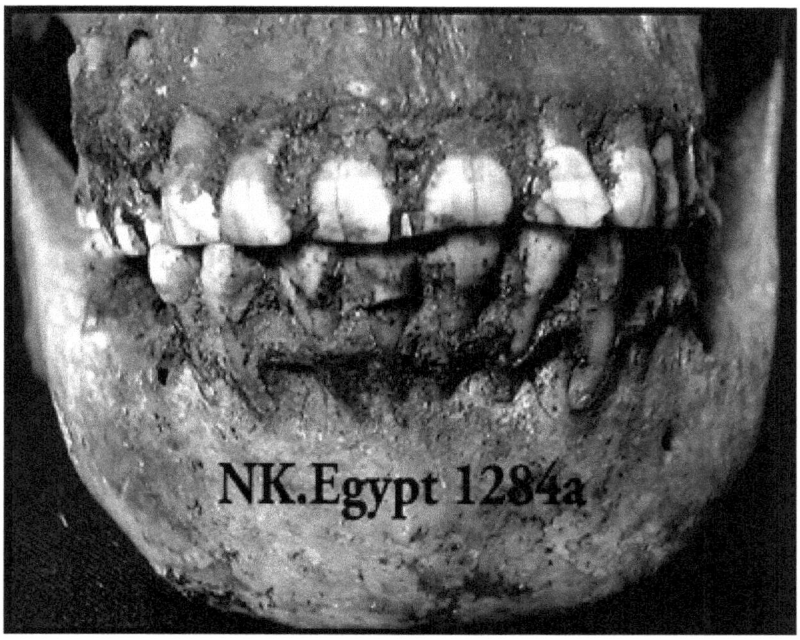

Fig 27b: ***Possible physico-chemical deposition of precipitated carbonates.***

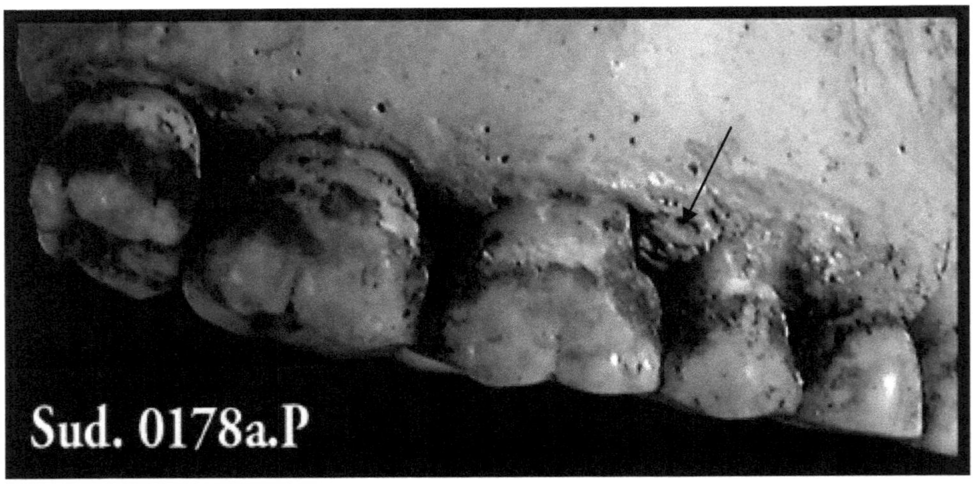

Fig. 28a: ***Evidence of mild horizontal periodontal disease***

Ref. 7.11.1

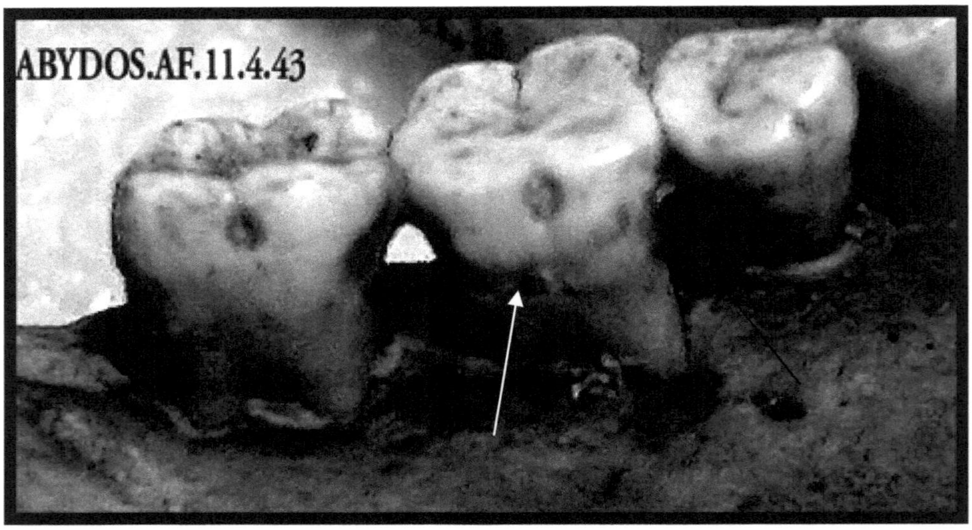

Fig. 28b: ***Evidence of horizontal periodontal disease.***

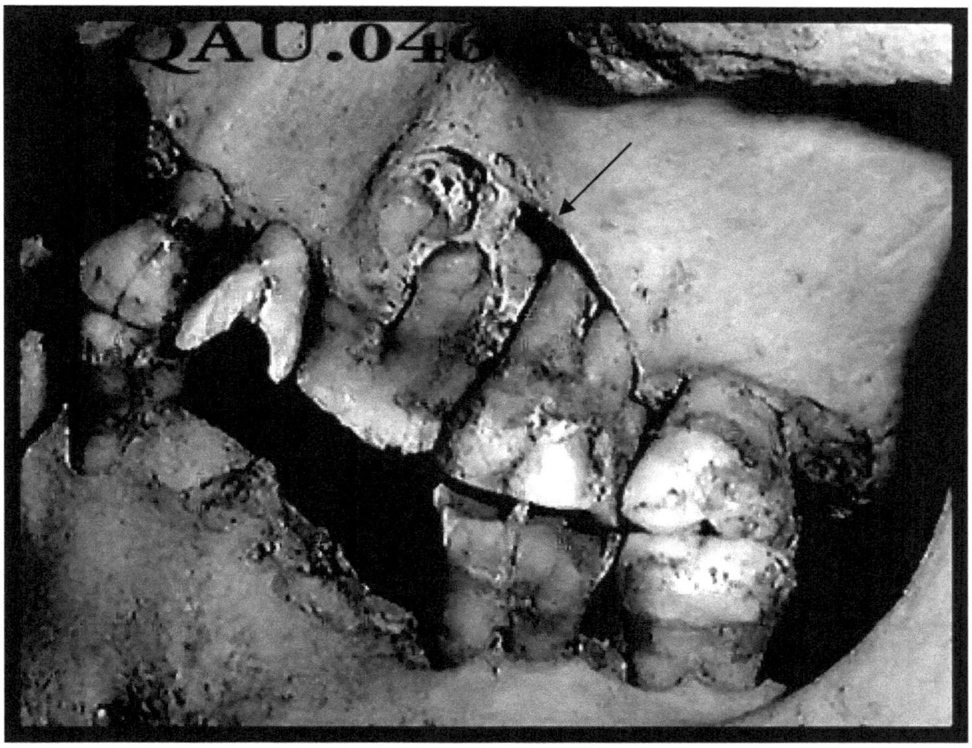

Fig. 29 a: ***Example of damage caused by vertical periodontal disease.***

Ref. 7.11.2

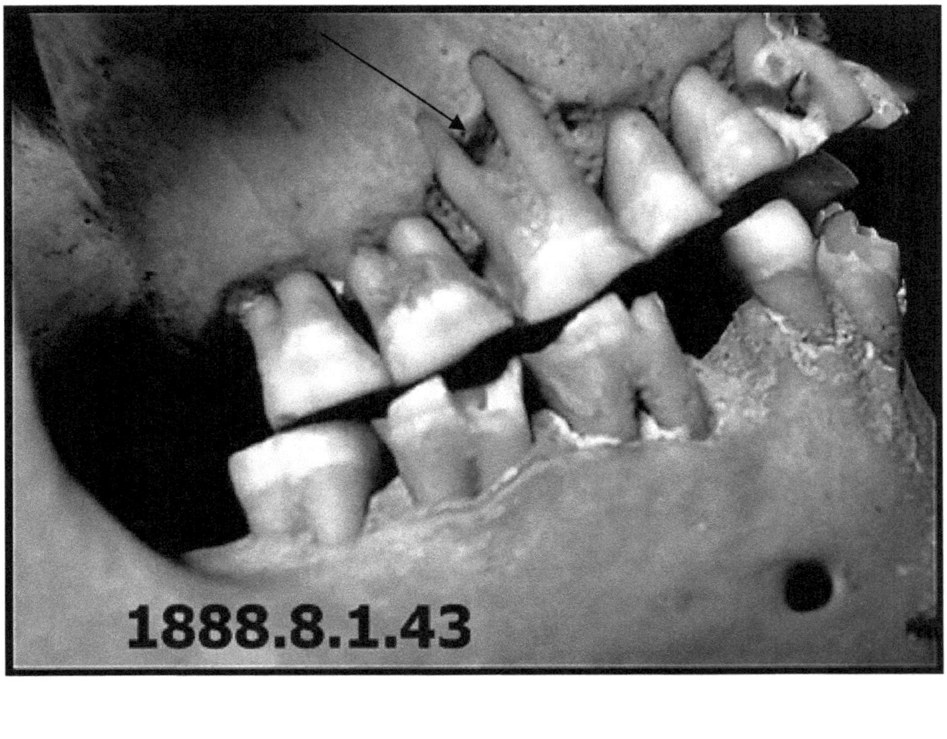

Fig. 29 b: ***Example of damage caused by vertical periodontal disease.***

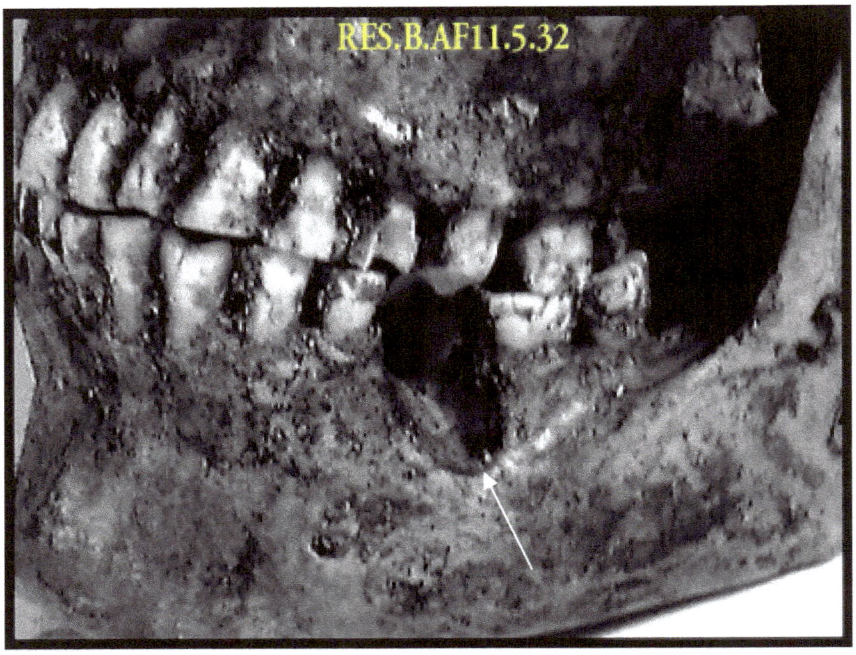

Fig. 30 a & b: ***Examples indicative of 'aggressive periodontitis'.***

Ref. 7.11.3

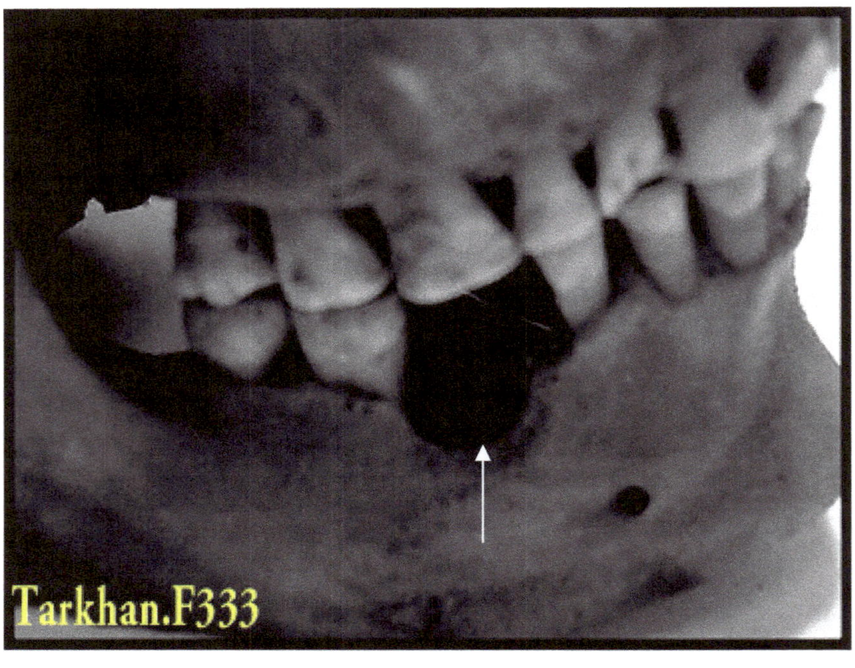

Fig 31 a, b &c: ***Examples of alveolar damage probably caused by apical infection.***

Ref.7.12

Fig 31 a.

Fig 31 b.

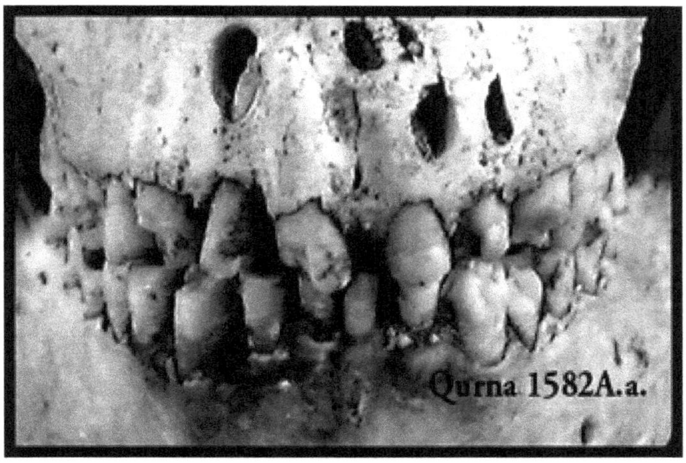

Fig 31 c.

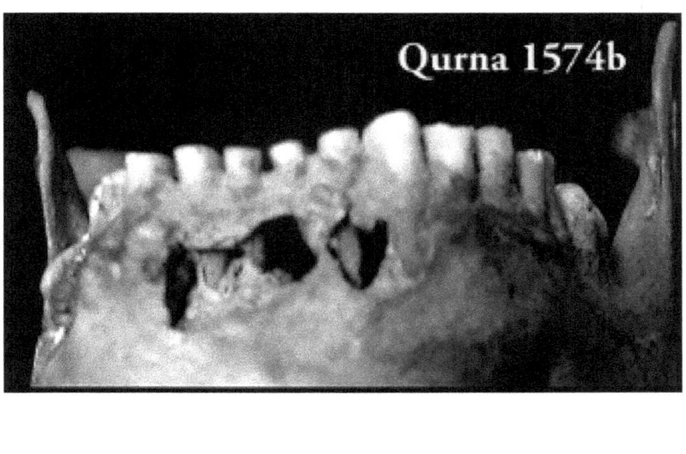

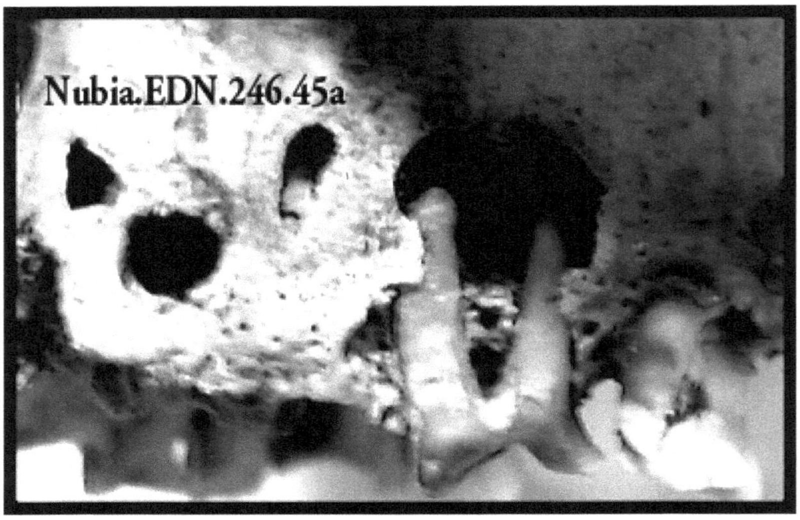

Fig. 31d: ***Alveolar fistula /6 caused by:-***

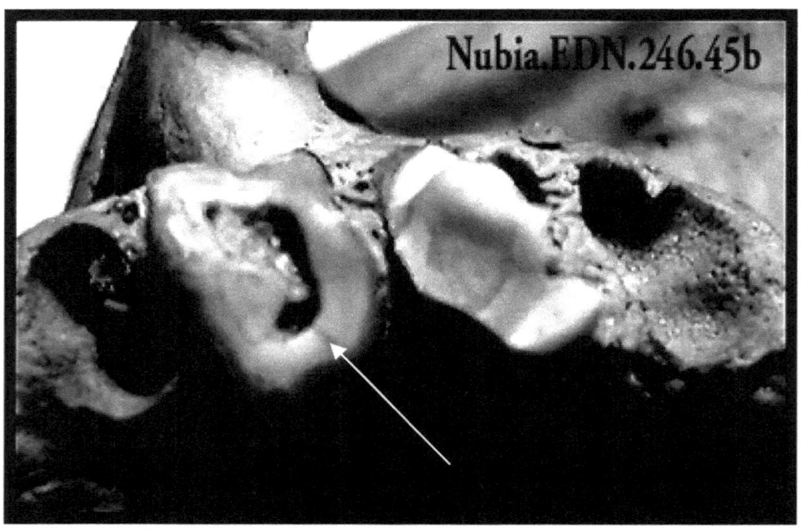

Fig. 31e: ***Exposure of the pulp through severe occlusal wear.***

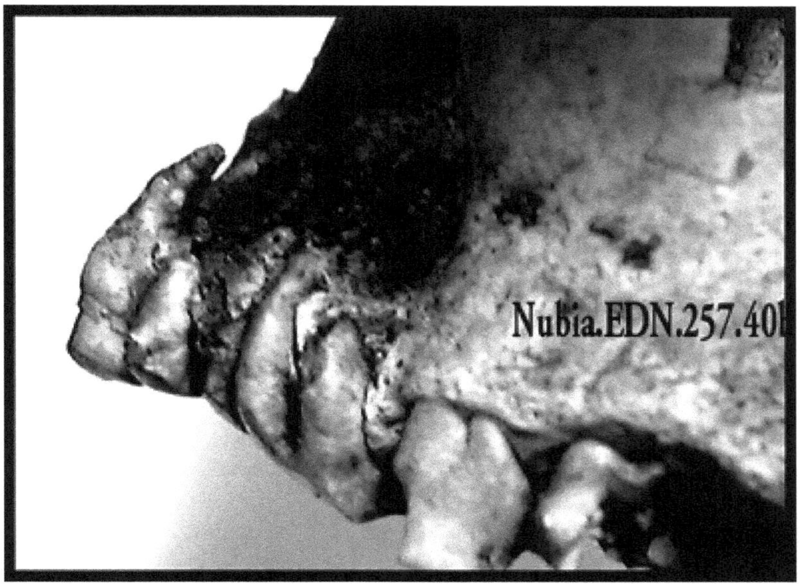

Fig. 31 f: ***Gross bone loss of unknown aetiology.***

Fig 32 a & b: ***Examples of alveolar damage in the Predynastic period suggestive of cysts.***

Ref. 7.13

Fig 32 a.

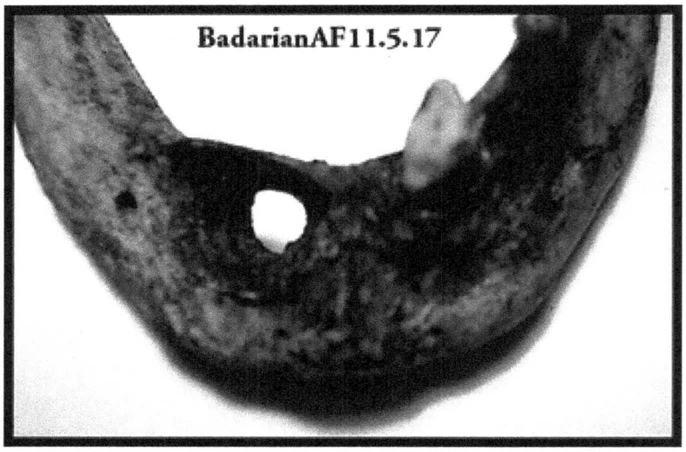

Fig 32 b.

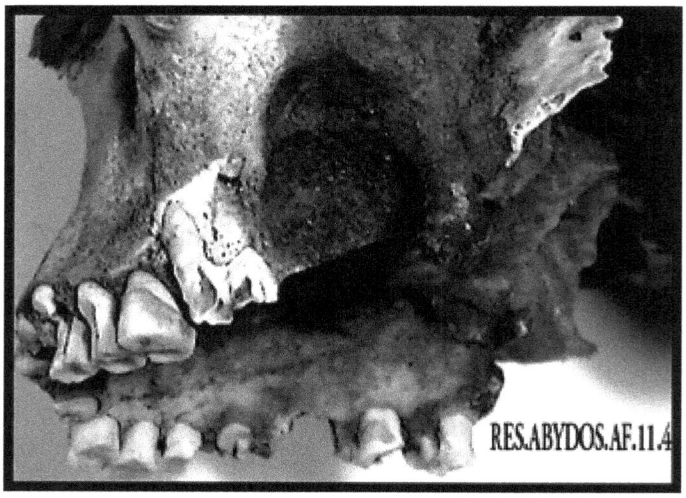

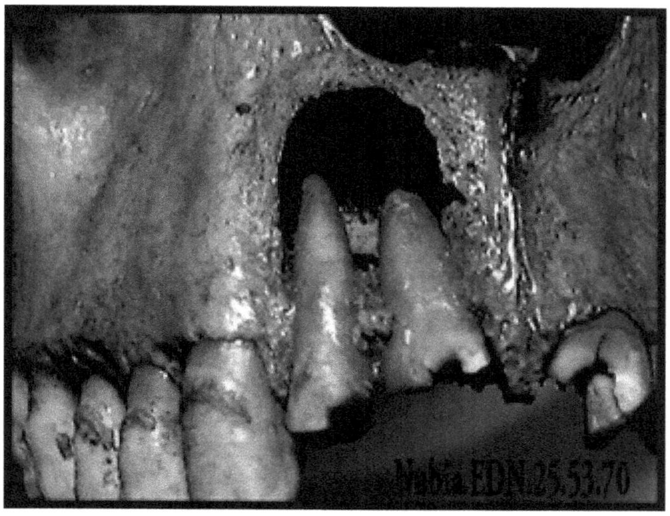

Fig 32 c: ***Example of probable cyst involving 21/.***

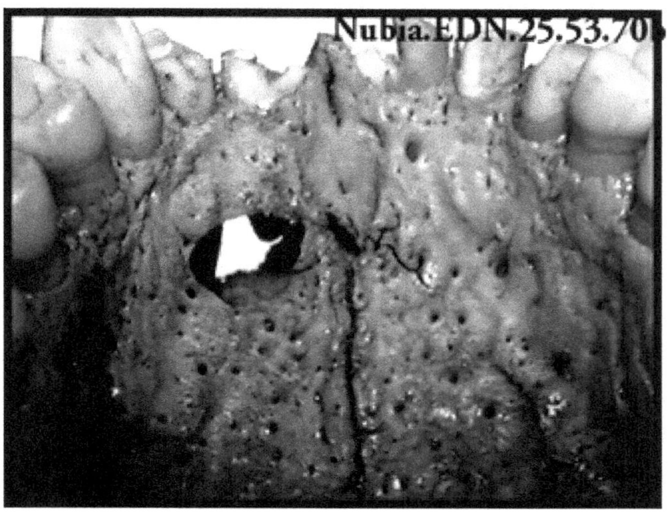

Fig 32 d: ***Palatal view of same lesion.***

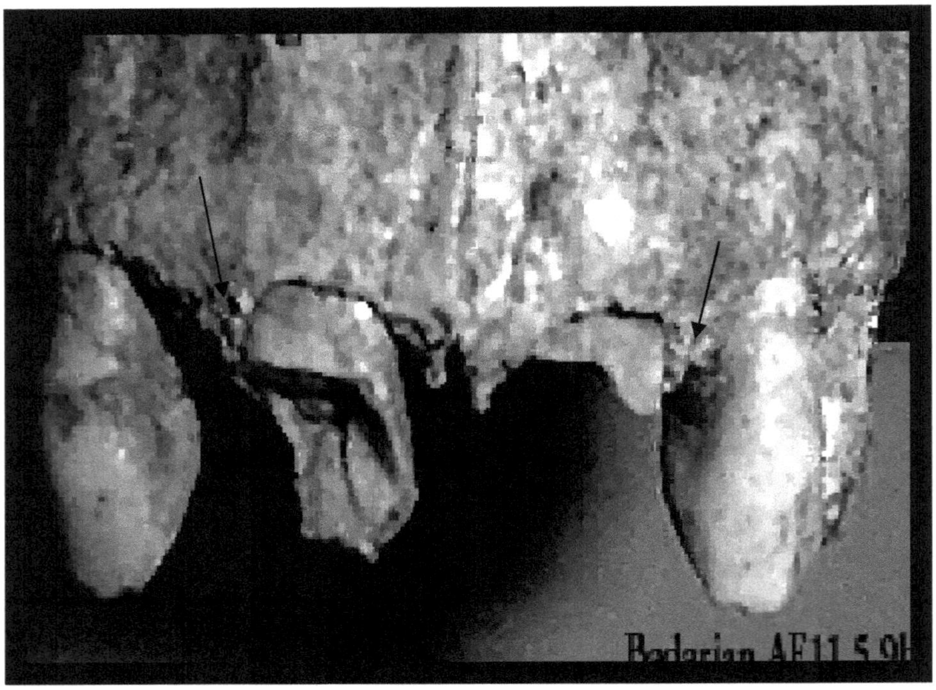

Fig. 33a: ***Hypodontia – absent 2/2***

Ref. 7.14.1

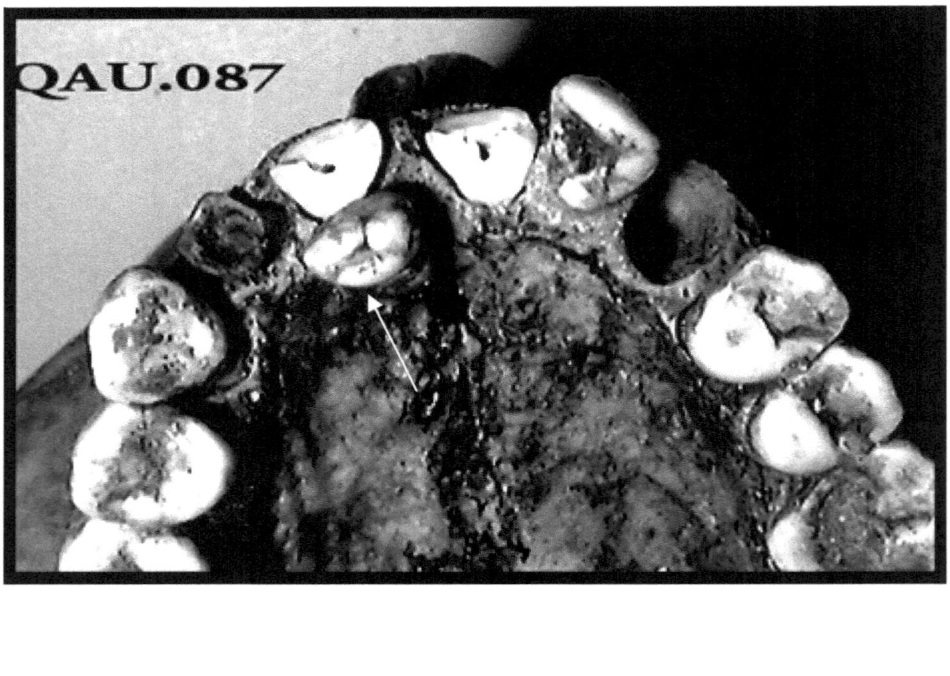

Fig. 33b: ***Hyperdontia – supplemental 1/***

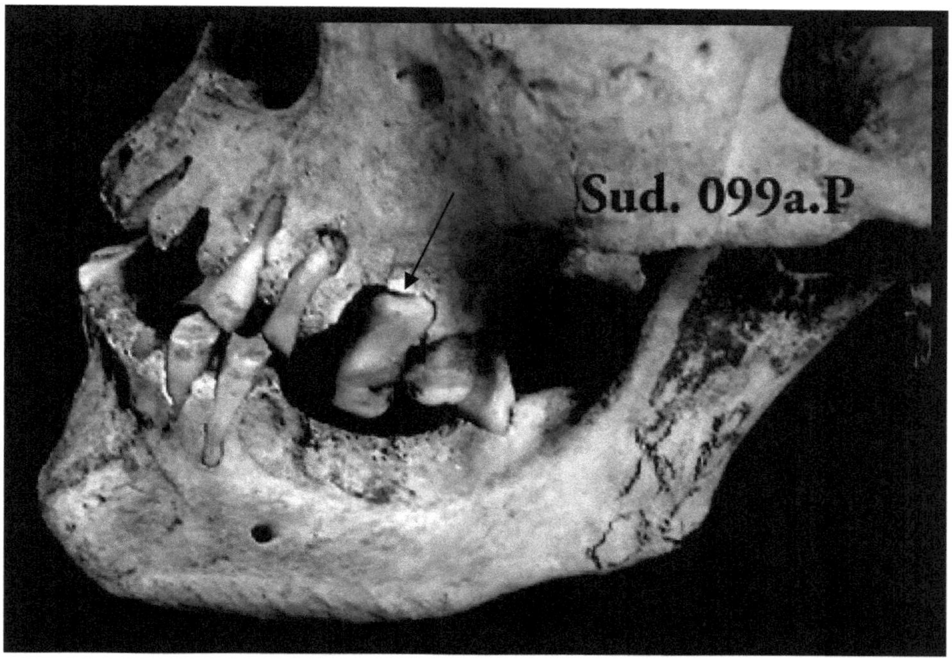

Fig.34a: ***Taurodont /6.***

Ref. 7.14.3

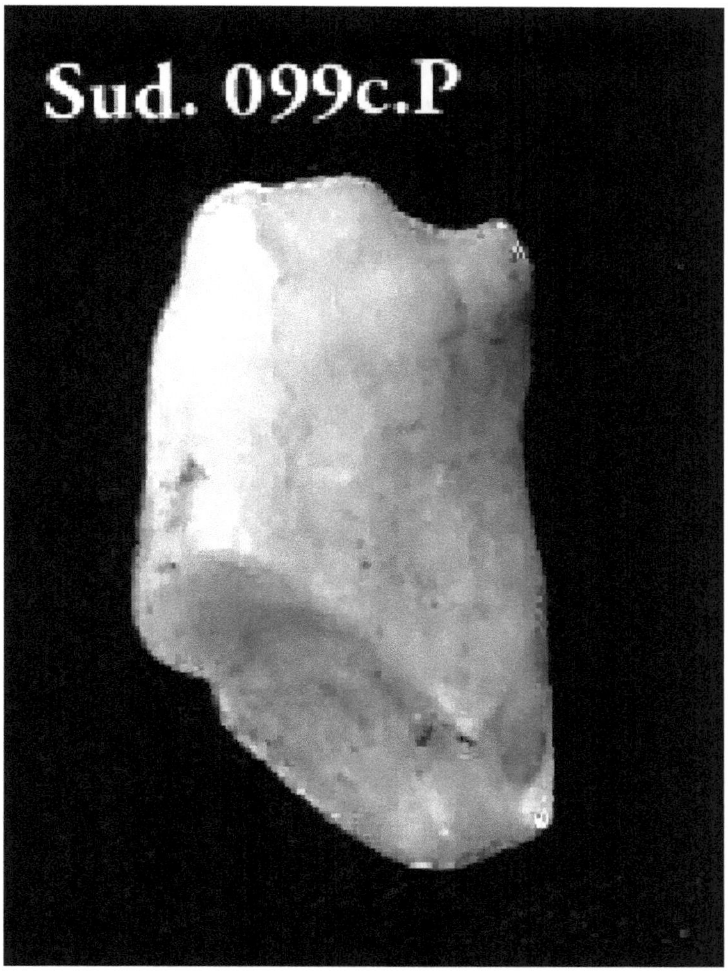

Fig.34b: ***Taurodont /6. Occlusal surface exhibiting oblique wear.***

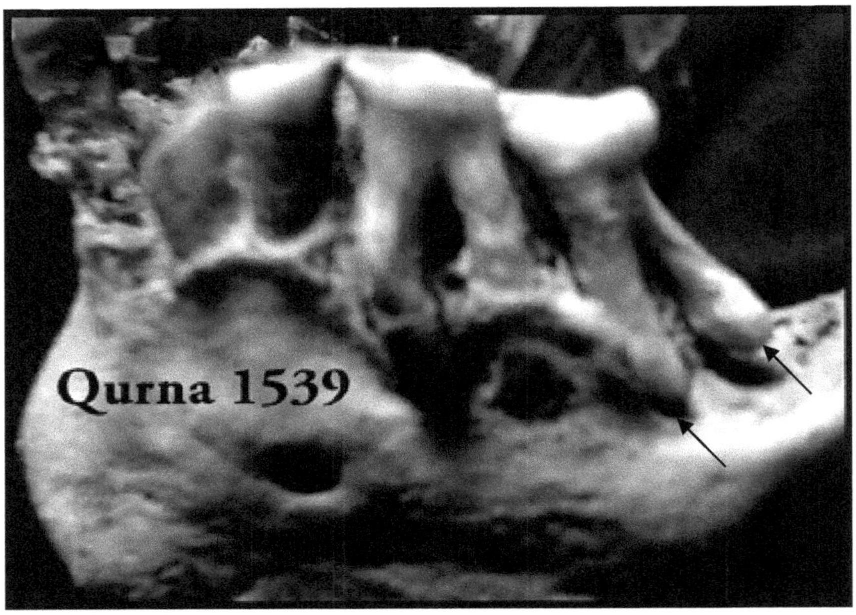

Fig.35: ***Hypercementosis of roots of LL7***

Ref. 7.14.4

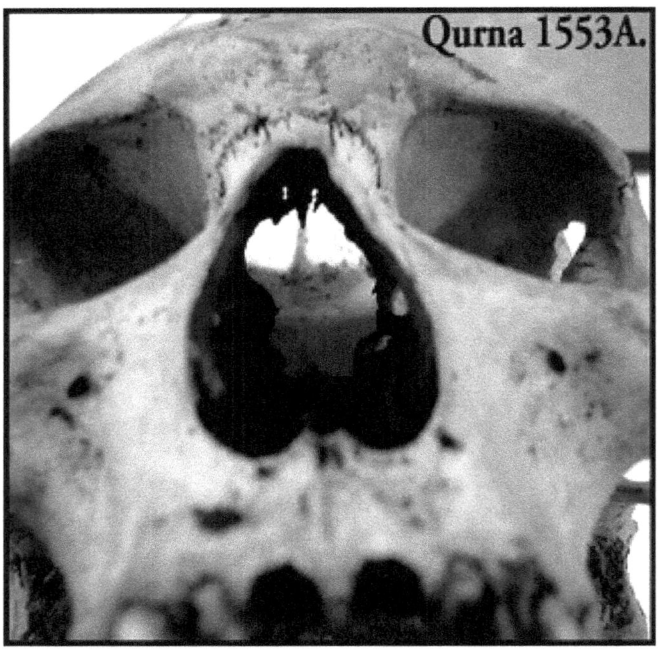

Fig.36a: ***Excerebration through the ethmoid with destruction of the turbinate bones.***

Ref. 7.15

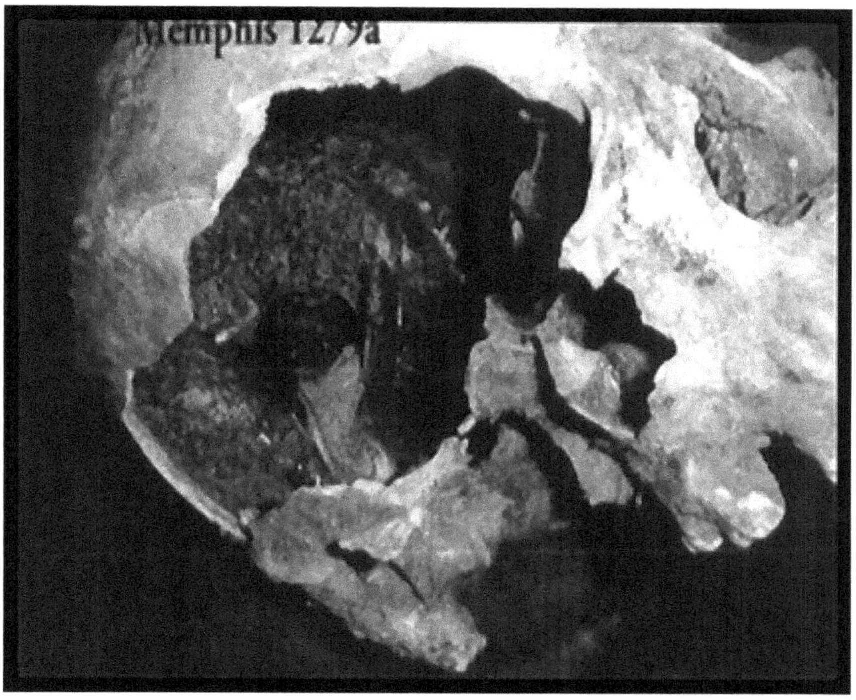

Fig.36b: ***Fractured skull showing resin which has been used for packing.***

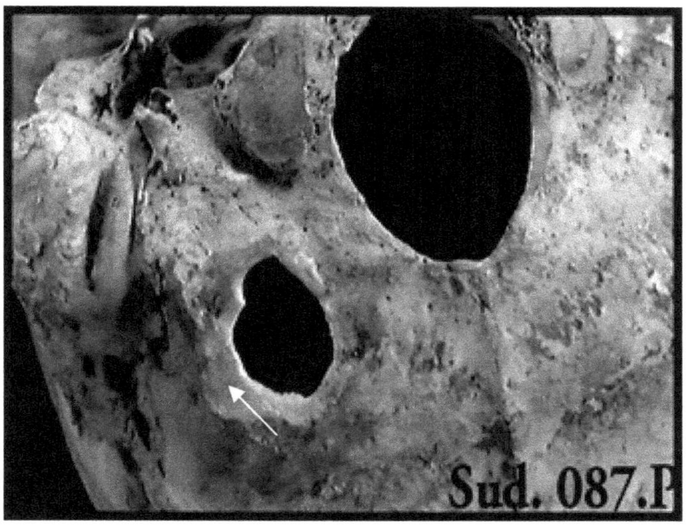

Fig.36c: ***Possible site for excerebration or filling skull with packing material?***

Ref. 7.15

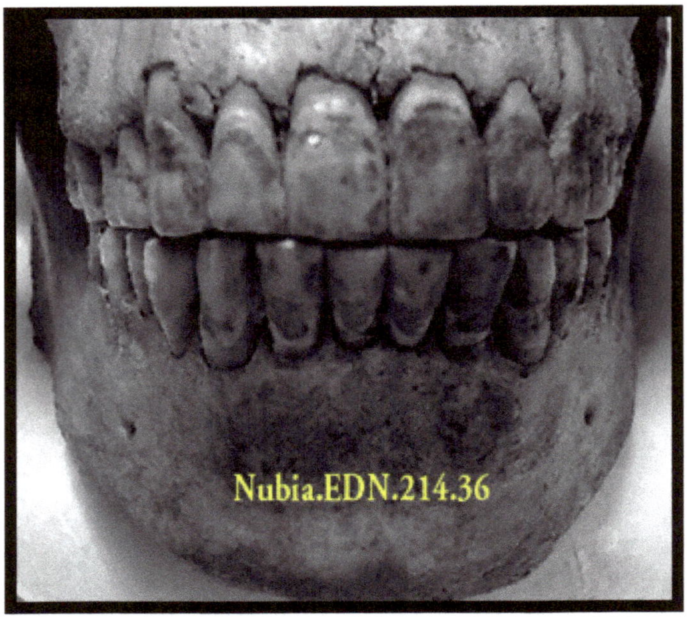

Fig. 37a: ***Class 1 labial relationship.***

Ref. 7.16

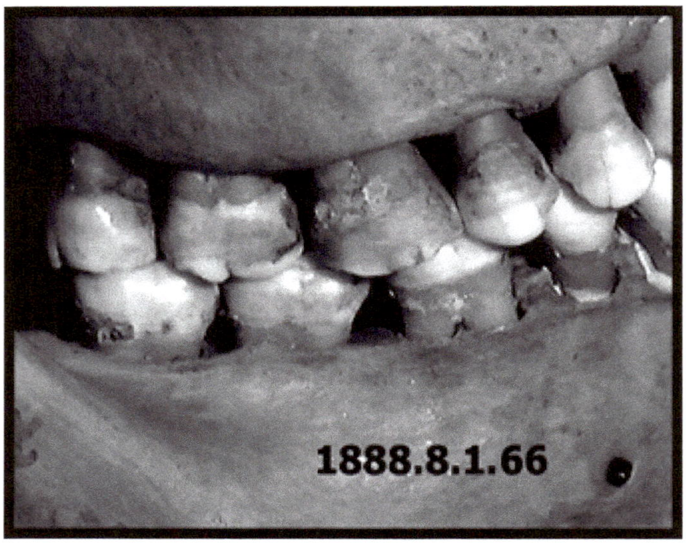

Fig. 37b: ***Class 1 buccal relationship.***

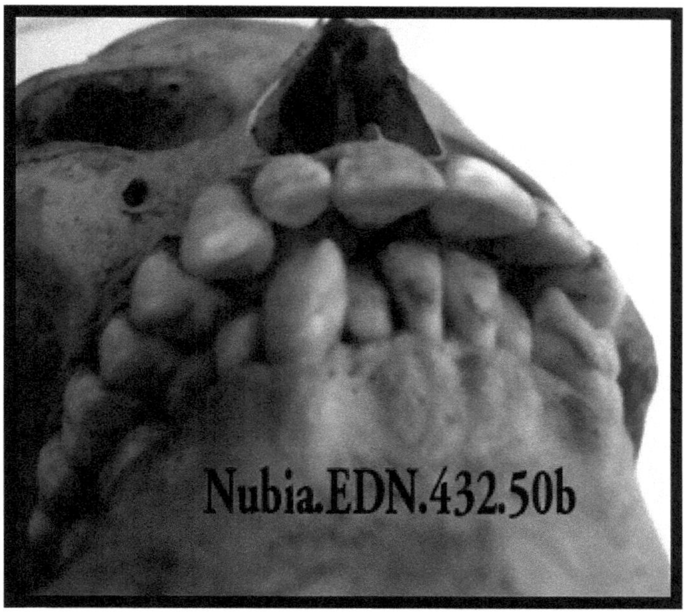

Fig 38a: ***Class 11 labial relationship.***

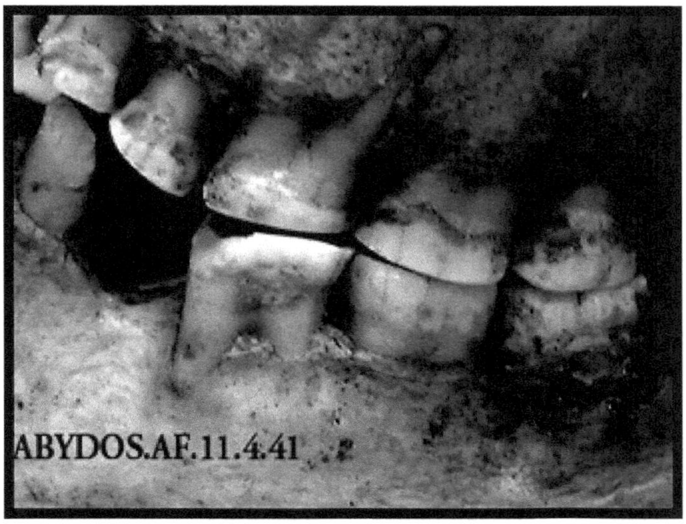

Fig 38b: ***Class 11 buccal relationship.***

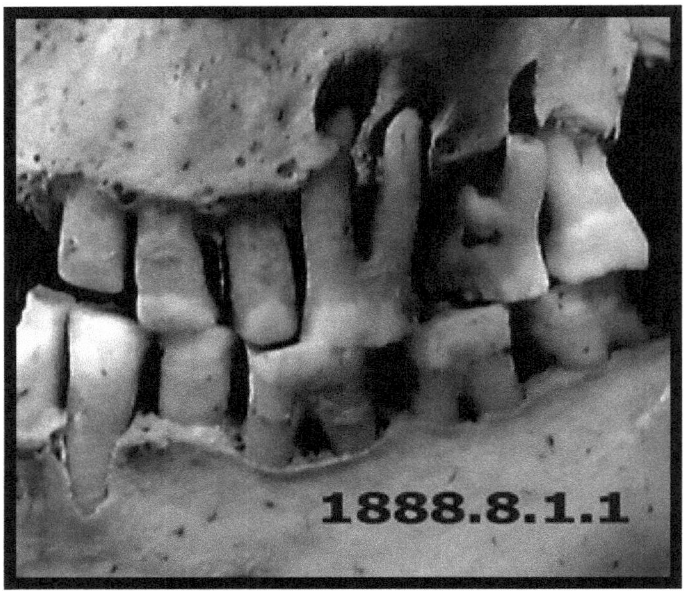

Fig 39a: ***Class 111 buccal relationship.***

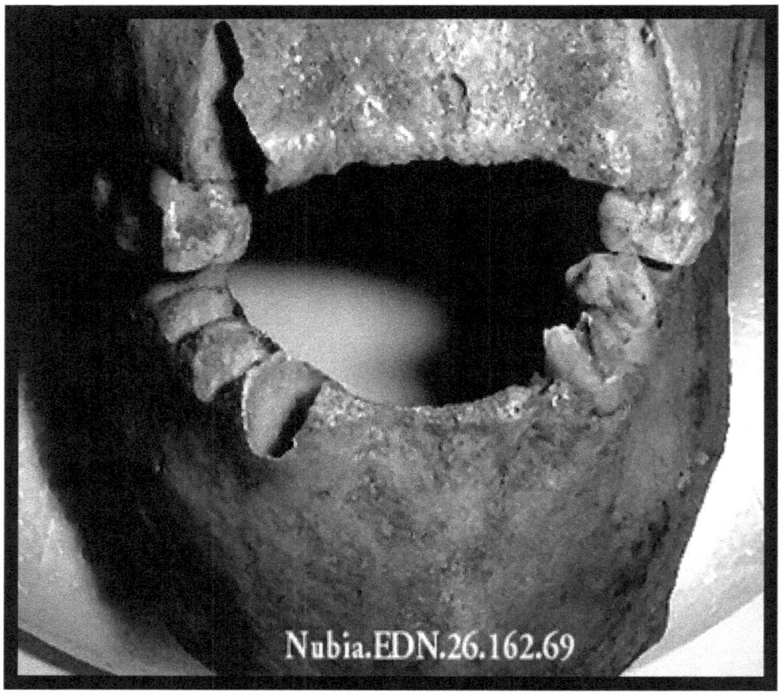

Fig 39b: ***Anterior open bite.***

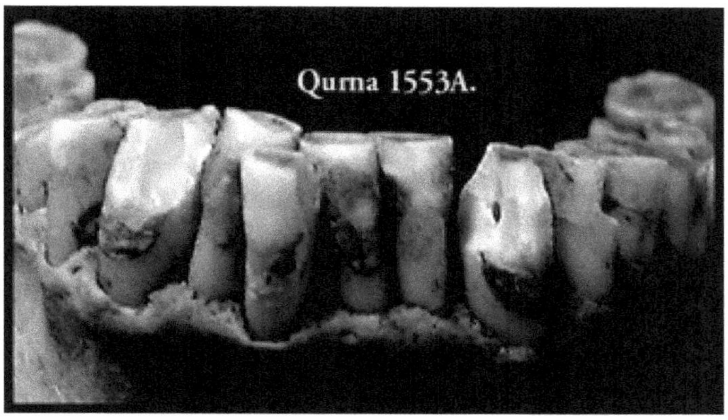

Fig. 40a: ***Crowding.***
Ref. 7.16

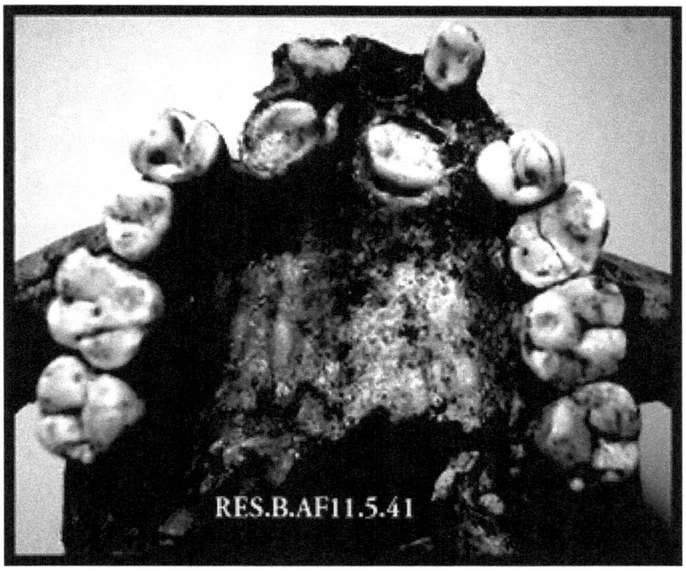

Fig. 40b: ***crowding and rotation of 5 / & retention of /E.***

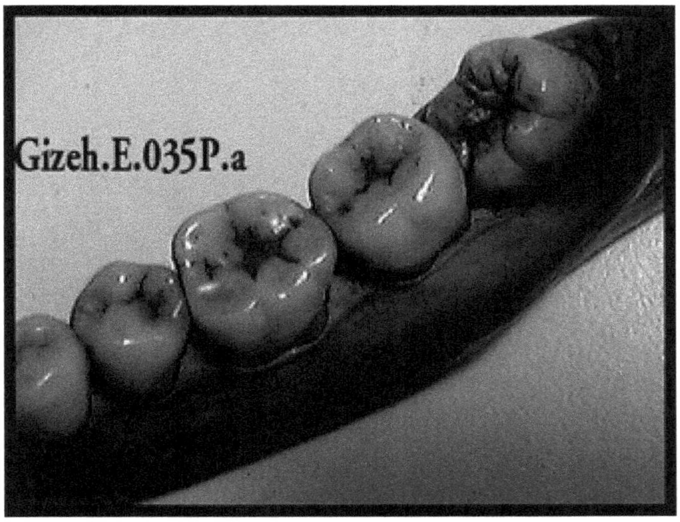

Fig. 40c: ***Mesio angular impaction of LL8 with no wear either on the occlusal or interproximal surfaces***

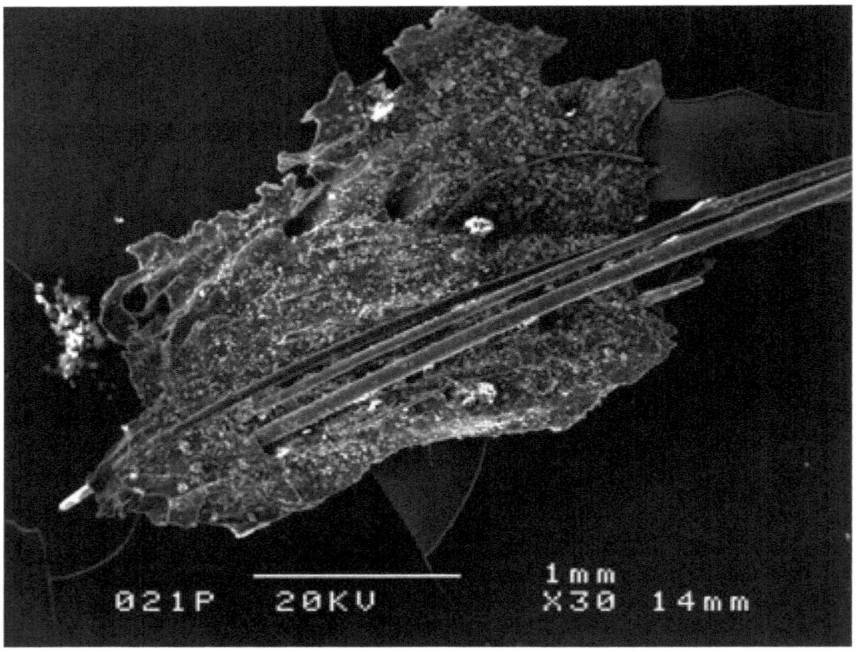

Fig. 41a: ***Flake of scalp with 3 hairs magnified x 30.***

Ref. 7.13

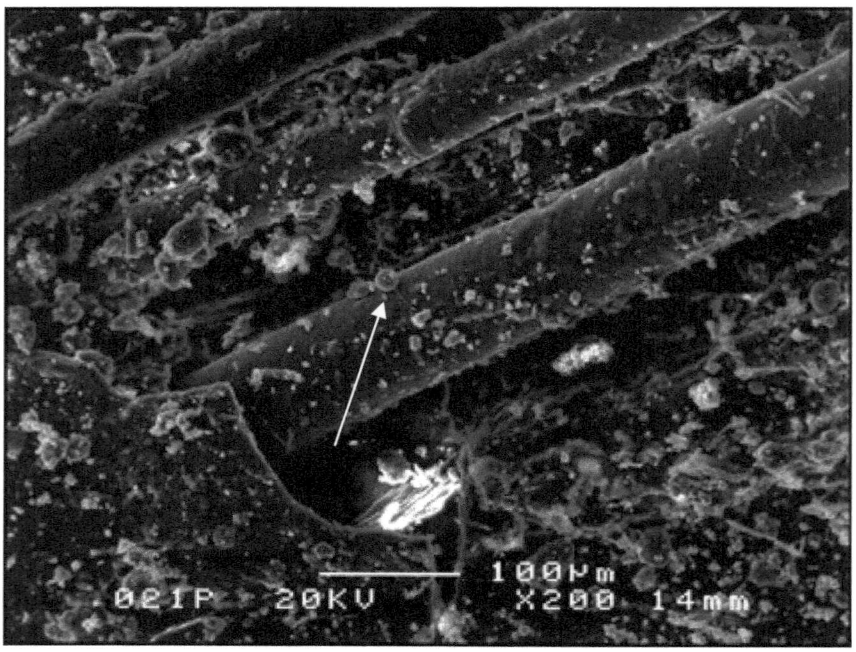

Fig. 41b : ***3 hairs magnified x 200 emerging from flake of scalp.***
Arrow pointing to pollen grain.

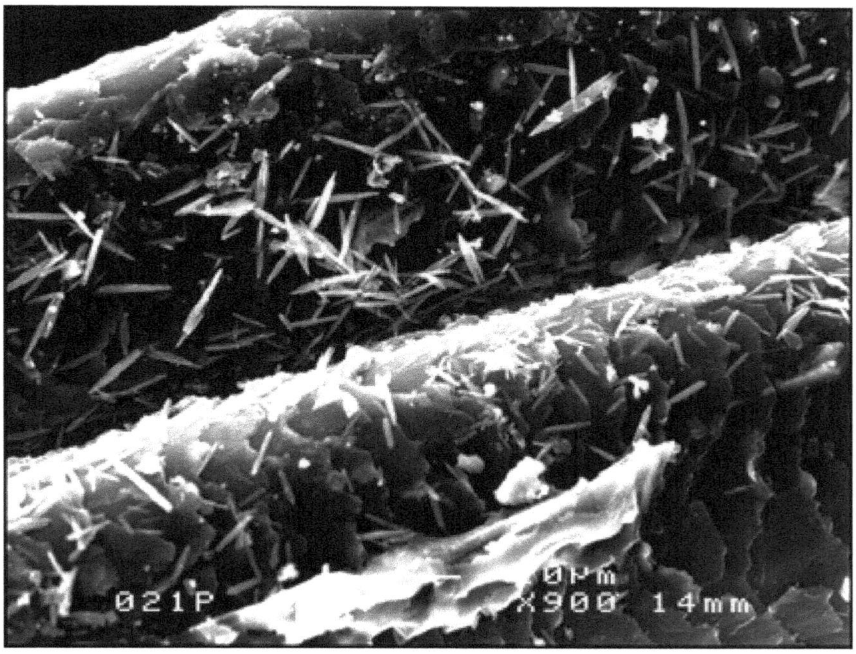

Fig. 41c : ***Hairs magnified x 900 showing scales on shaft and crystals of gypsum.***

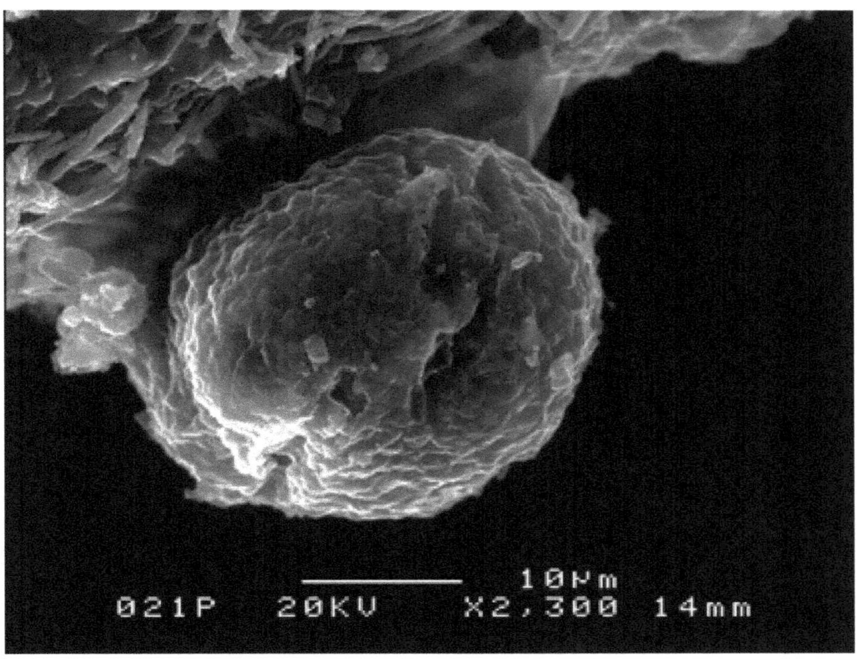

Fig. 41d: ***Pollen grain on hair magnified x 2300.***

References

ABBIE, A.A. 1950. Closure of cranial articulations in the skull of the Australian Aborigine. *Journal of Anatomy*. **84**, 1-12.

ALEXANDERSEN V. 1967. The pathology of the jaws and the temporomandibular joint. *In:* D. BROTHWELL and A.T. SANDISON, Eds., *Diseases in Antiquity*. Charles C. Thomas, Springfield, Illinois, U.S.A., 551-595.

ARMELAGOS, G.J. and MILLS, J.O. 1993. Palaeopathology as science: the contribution to Egyptology. *In:* W.V.DAVIES and R.WALKER. Eds., *Biological Anthropology and the study of Ancient Egypt.* British Museum Press, London, 1-18.

AYKROYD, W.R., 1967. The Story of Sugar. *Quadrangle Books.* Chicago.

BAINES, J. and MALEK, J. 1958. The Historical Background. *An Atlas of Ancient Egypt.* Phaidon Press Ltd., Oxford, 30-35.

BEGG, P.R. 1954[1]. Stone Age Man's Dentition. *American Journal of Orthodontics,* **40,** 373-383.

BEGG, P.R. 1954 [2]. Stone Age Man's Dentition. 298-312.

BEGG, P.R. 1954 [3]. Stone Age Man's Dentition. 462-475.

BEGG, P.R. and **KESLING, P.C.** 1977. Correct Occlusion, the basis for Orthodontics. *Begg Orthodontics, Theory and Technique.* W.B. Saunders Co., Philadelphia, 7-40.

BENNIKE, P. and FREDEBO, L. 1986. Dental treatment in the Stone Age. *Bulletin of the History of Dentistry.* **34,** no.2, 81-87.

BLACKWOOD, H.J.J., 1963. Arthritis of the mandibular joint. *British Dental Journal.* **115,** no. 8, 317-326.

BOYD, W. 1958. Diseases of the Bone. *Pathology for the Physician.* Kimpton. London, 809.

BRABANT, H. 1967. Paleostomatology. *In:* D. BROTHWELL and A.T. SANDISON, Eds., *Diseases in Antiquity*. Charles C. Thomas, Springfield, Illinois, U.S.A., 435-470.

BRADFORD, J. 1958. Food and the periodontal diseases. Nutrition and teeth. *Proceedings of the Nutrition Society.* **18,** 75-79.

BREASTED, J.H. 1930. *The Edwin Smith Surgical Papyrus.* The University of Chicago Oriental Institute Publications. **1.**

BREWER, D.J., REDFORD, D.B. and REDFORD, S. 1994 . Fruits. *Domestic plants and animals. The Egyptian origins.* Aris and Phillips Ltd., Warminster, England. 47-64.

BREWER, D.J., REDFORD, D.B. and REDFORD, S. 1994 [2]. The birds and the bees. 120-129.

BROTHWELL, D., CARBONELL, V.M. and GOOSE, D.H. 1963. Congenital absence of teeth in human populations. *Dental Anthropology.* Pergamon Press. Oxford.

BROTHWELL, D. 1963. Macroscopic dental pathology of humans. *In:* D. BROTHWELL, Ed., *Dental Anthropology.* Pergamon Press, Oxford.

BROTHWELL, D. 1967[2]. Evidence for neoplasms. *In:* D. BROTHWELL and A.T. SANDISON., Eds., *Diseases in Antiquity.* Charles C. Thomas. Springfield, Illinois, U.S.A. 320-346.

BROTHWELL, D. 1967. The bio-cultural background to disease. *In:* D. BROTHWELL and A.T. SANDISON., Eds., *Diseases in Antiquity.* Charles C. Thomas, Springfield, Illinois, U.S.A. Ch. 5.

BROTHWELL, D. 1994. *Digging up Bones.* British Museum (Natural History). Cornell University Press, Ithaca. 154.

CECIL, R.L. and LOEB, R.F. 1960 . Bacterial Diseases. *A Textbook of Medicine.* W.B.Saunders Co., Philadelphia and London. 194-201.

CECIL, R.L. and LOEB, R.F. 1960 [2]. Deficiency Diseases. *A Textbook of Medicine.* W.B.Saunders Co., Philadelphia and London. 537-539.

CHANDRA, R.K. 1993. Trace elements and immune responses. *Trace elements in nutrition of children.* Nestles Nutrition Series, Raven Press, New York. **11**.32.

CLARKE, N.G., CAREY, S.E., SRIKANDI, W., HIRSCH, R.S. and LEPPARD, P.I. 1986. Periodontal disease in ancient populations. *American Journal of Physical Anthropology.* **71**. 173-183.

CLARKE, N.G. and HIRSCH, R.S. 1991. Physiological, pulpal and periodontal factors influencing alveolar bone. *In:* M.A. KELLEY and C.S. LARSEN, Eds., *.Advances in dental anthropology.* Wiley-Liss NY., 241-266.

CORNWELL, J. 2000. *Trick or Treatment.* Times Supplement. Saturday 24/12/2000.

COTTONE, J.A. and STANDISH, S.M. 1981. Outline of forensic dentistry. *Year Book Medical Publishers,* Chicago.

CRAIG, P. and DAVEY, J. 2001. Possible evidence for ancient Egyptian orthodontics. *Egyptian Archaeology. Bulletin of the Egypt Exploration Society.* London. **18**. 37-39.

DARBY, W.J., GHALIOUNGUI, P and GRIVETTI, L. 1977. *Food: The Gift of Osiris.* Academic Press. London and New York. **1 and 2.**

DAVID, A.R. 1982. The Predynastic and Early Dynastic Communities. *The Ancient Egyptians.* Routledge, London. 10-41.

DAVID, R. and TAPP, E., Eds., 1984. *Evidence Embalmed.* Manchester University Press, Manchester. 23.

DAVID, A.R. and DAVID, A.E. 1992. *A Biographical Dictionary of Ancient Egypt.* Seaby Ltd., London.

DAVID, A.R. 1992. Natsef-Amun's Life as a Priest. *In: A.R.. DAVID and E. TAPP,* Eds., *A Mummy's Tale.* Michael O'Mara Books Ltd., London. 65-80.

DAVID, R. 1999. *The Experience of Ancient Egypt.* Routledge. London.

DAWSON, W. 1929. *Magician and Leech.* Methuen, London.

DAWSON, W. 1967. The Egyptian Medical Papyri. *In:* D. BROTHWELL and A.T. SANDISON., Eds., *Diseases in Antiquity.* Charles C. Thomas, Publisher. U.S.A. 98-111.

DAYAGI-MENDELS, M. 1989. *Perfumes and Cosmetics in the Ancient World.* Catalogue 305. Israel Museum.

DIXON, D.M. 1972. Masticatories in Ancient Egypt. *Journal of Human Evolution,* **1**. 433-449.

DOXIADIS, E. 1995. *The Mysterious Fayum Portraits.* Thames and Hudson. London.

DROWER, M.S. 1985. Most Ancient Egypt. *Flinders Petrie.* Gollanz. London. 249-274.

DUNAND, F. and LICHTENBERG ,R. 1991. *Mummies: a Journey through Eternity.* New Horizons. Thames and Hudson. London.

DUNLOP, D. and ALSTEAD, S. 1966 Nutritional Disorders. *Textbook of Medical Treatment.* E. and S. Livingstone, Edinburgh & London. 384.

EBBELL, B. 1937. *The Papyrus Ebers.* Levin and Munksgaard, Ejjnar Munksgaard. Copenhagen.

EDGAR, W.M. 1993. Extrinsic and Intrinsic Sugars: A review of recent UK recommendations on diet and caries. *Caries Research No 27.* Supplement **1**, 64-67.

ELASHOFF, J.D. 2000. *JD. nQuery adviser 4.0 User's Guide.* Los Angeles, CA.

EMERY, W.B. 1961. *Archaic Egypt.* Penguin Books, London. 236-240.

ENCYCLOPAEDIA BRITANNICA C.D. 2002. *Piltdown Man. Dawson's Dawn Man.*

ENCYCLOPAEDIA BRITANNICA C.D. 2002. *Gypisol.*

ENCYCLOPAEDIA BRITANNICA C.D. 2002. *Athenaeus.*

ESTES, J.WORTH. 1993. A glossary of Drug Substances. *The Medical Skills of Ancient Egypt.* Science History Publications. U.S.A. 139-157.

EVERSOLE, L.R., PAPPAS, J.R. and GRAHAM, R. 1985. Dental Occlusal Wear and degenerative disease of the temporomandibular joint: a correlational study utilizing skeletal material from a contemporary population. *Journal of Oral Rehabilitation.* **12,** 401-406.

EYRE, C.J. 1999. Village Economy in Pharaonic Egypt. *In:* A.K. BOWMAN and E. ROGAN, Eds., *Agriculture in Egypt.* British Academy by Oxford Univ. Press. 33-61.

FERGUSON, M.W.J. 1978. Cleft palate past and present. Palaeopathology News Letter. **24.** 5-8.

FILER, J. 1995. Dental health. *Disease.* British Museum Press. 101.

FOLEY, R.A. 1999. The Duckworth Osteological Collection. Internet information.

GARDINER, A. 1961. *Egypt of the Pharaohs.* Oxford University Press. 19-40.

GARNER, V., HALL, A. and HALL, K. 1997. Application of Micro-Scale Sealed Vessel Pyrolysis-Gas

Chromatography-Mass Spectrometry. (MSSV-GC-MS) *Chromatography and analysis.* **9,** July/August.

GHALIOUNGHUI, P. 1963. *Magic and Medical Science in Ancient Egypt.* Hodder and Stoughton, London.

GHALIOUNGHUI, P. 1980. Medicine in Ancient Egypt. *In:* J.E. HARRIS and E.F. WENTE, Eds., *An X-Ray Atlas of the Royal Mummies.* University of Chicago Press.

GHALIOUNGHUI, P. 1983. ***The physicians of Pharaonic Egypt.*** National Library of Medicine. US. Al-Ahram Center for Scientific Translations.

GHALIOUNGHUI, P. 1987. *Translation of Ebers Papyrus.* By kind permission of W. Benson Harer, San Bernardino, Ca. U.S.A.

GILKEY, F. 1978. Cleft lip and palate in palaeopathological material. *Palaeopathology News Letter.* **24,** 8-11.

GILLES, H.M. 1989. Haemoglobinopathies and Clucose-6-phosphate dehydrogenase deficiency. *In:* ADAMS and MAEGRAITH, Eds., *Clinical Tropical Diseases.* 9th Edition. Blackwell Scientific Publications. 115-127.

GOLDEN, M.H.N. 1996 Severe Malnutrition. *In:* WEATHERALL, LEDINGTON and WORRELL, Eds., *Oxford Textbook of Medicine.* 3rd Edition. Oxford University Press. **1,** 1278-1296.

GRAY, R.J.M., DAVIES, S.J. and QUAYLE, A.A. 1994. A Clinical approach to temporomandibular disorders. Should dentists treat headache? *British Dental Journal.* **177,** 255-259.

GRILLETTO, R. 1973. Caries and Dental Attrition in the Early Egyptians as seen in the Turin Collections. *In*: D.R. BROTHWELL, and B.A. CHIARELLI, Eds., *Population Biology of the Ancient Egyptians.* Academic Press. London & N.Y. 325-331.

GUSTAFSON, G. 1950. Age determination on teeth. *Journal of the American Dental Association,* **41,** 45-54.

HAMDAN, A.E. and ROCK, W.P. 2001. Cephalometric Norms in an Arabic Population. *Journal of Orthodontics.* **28,** No.4, 297-300.

HARER, BENSON W. 1993. Health in Pharaonic Egypt. *In:* W.V. DAVIES and R. WALKER, Eds., *Biological Anthropology and the study of Ancient Egypt.* British Museum Press, London, 19-23.

HARRIS, J.E., ISKANDER, Z. and FARID, S. 1975. Restorative Dentistry in Ancient Egypt: an Archaeological Fact. *The Journal of the Michigan Dental Association.* **57.** 401-404.

HARRIS, J.E., KOALSKI, C. and WALKER, G.F. 1980. Craniofacial variation in the Royal Mummies, *In:* J.E. HARRIS and E.F. WENTE, Eds., *An X-Ray Atlas of the Royal Mummies.* University of Chicago Press, 346-357.

HARRIS, J.E., STOREY, A.T. and PONITZ, P.V. 1980. Dental Disease in the Royal Mummies. *In:* J.E. HARRIS and E.F. WENTE, Eds., *An X-Ray Atlas of the Royal Mummies.* University of Chicago Press, 328-346.

HART, G. 1986. *Egyptian gods and goddesses.* Routledge and Kegan Paul Inc. London.

HARTNADY, P. and ROSE, J.C. 1991. Archaic Lower Pecos tooth-loss patterns. *In:* M.A. KELLEY and C.S. LARSEN, Eds., *.Advances in dental anthropology.* Wiley-Liss NY., 267-278.

HAYES, W.C. 1964. Neolithic and Chalcolithic communities in Northern Egypt. *In*: K.C. SEELE, Ed., *Most Ancient Egypt,* Chicago University Press. 91-146.

HERODOTUS. 1972. *The Histories.* Translated by A. de Selincourt. Penguin Classics. London. Book 2.

HERTZ, J.H. 1976. *Pentateuch and Haftorahs.* 2nd edition. Soncino Press, London, 614.

HILDEBOLT, C.F. and MOLNAR, S., 1991. Measurement and description of periodontal disease in anthropological studies. *In:* M.A. KELLEY and C.S. LARSEN, Eds., *Advances in dental anthropology.* Wiley-Liss, NY., 225-240.

HILLSON, S. 1996. *Dental Anthropology.* Cambridge University Press.

HINTON, R.J. 1982. Differences in interproximal and occlusal wear among prehistoric Tennessee Indians. *American Journal of Physical Anthropology,* **57,** 103-115.

HODGE, P.J., TEAGUE, P.W., WRIGHT, A.F. and KINANE, D.F. 2000. Clinical and genetic analysis of a large North European Caucasian family affected by early-onset periodontitis. *Journal of Dental Research.* **79,** No.3, 857-863.

HODGES, D. 1991. Temporomandibular osteoarthritis in a British skeletal population. *American Journal of Physical Anthropology.* **85,** 367-377.

HUNT, A.S. 1911. Rylands Greek Papyrus. *Cat. Greek Papyri, John Rylands Library. Medical Receipts,* **29(a).**

HURRY, J.B. 2000. *Imhotep. The Vizier and Physician of King Zoser and afterwards the Egyptian God of Medicine.* Sandpiper Books. Oxford University Press, USA.

JOHNSTON, T.B., DAVIES, D.V. and DAVIES, F. 1958. *Osteology.* Gray's Anatomy. Longmans, Green & Co. London. 264-355.

KEEN, J.A. 1950. Study of the differences between male and female skulls. *American Journal of Physical Anthropology.* **8**, 65-79.

KELLEY, M.A., LEVESQUE, D.R., and WEIDL, E. 1991. Contrasting patterns of dental disease in five early northern Chilean groups. *In:* M.A. KELLEY and C.S. LARSEN, Eds., *.Advances in dental anthropology.* Wiley-Liss, NY., 203-213.

KERR, N.W. and RINGROSE, T.J. 1998. Factors affecting the lifespan of the human dentition in Britain prior to the seventeenth century. *British Dental Journal.* **184.** No..5. 242-246.

LANG, N., BARTLOD, P.M., CULLINAN, M., JEFFCOAT, M., MOMBELLI, A., MURAKAMI, S., PAGE, R., PAPAPANOU, P., TONETTI, M. and VAN DYKE, T. 1999. Consensus report: Aggressive periodontitis. *Annals of Periodontology.* **4,** 53.

LARSON, C.S., SHAVIT, R. and GRIFFIN, M.C. 1991. Dental caries evidence for dietary change: an archaeological context. *In:* M.A. KELLEY and C.S. LARSEN, Eds., *.Advances in dental anthropology.* Wiley-Liss, NY., 179-203.

LAST, R.J. 1981. *Anatomy. Regional and applied.* Churchill and Livingstone. Edinburgh.

LEEK, F.F. 1966. Observation of Dental Pathology in Ancient Egyptian Skulls. *Journal of Egyptian Archaeology.* **52,** 59-64.

LEEK, F.F. 1967a. The Practice of Dentistry in Ancient Egypt. *Journal of Egyptian Archaeology.* **53,** 51-58.

LEEK, F.F. 1967b. Reputed Early Egyptian Dental Operation, An Appraisal. *In:* D. BROTHWELL AND A.T. SANDISON., Eds., *Diseases in Antiquity.* Charles C. Thomas, Springfield, Illinois, U.S.A. 702-705.

LEEK, F.F. 1969. The Problem of Brain Removal during Embalming by the Ancient Egyptians. *Journal of Egyptian Archaeology.* **55,** 112-116.

LEEK, F.F. 1972. Teeth and Bread in Ancient Egypt. *Journal of Egyptian Archaeology.* **58,** 126-132.

LEEK, F.F. 1973. Bite, attrition and associated oral conditions as seen in ancient Egyptian skulls. *In:* D.R. BROTHWELL and B.A. CHIARELLI, Eds., *Population Biology of the Ancient Egyptians.* Academic Press. London & N.Y. 290-295.

LEEK, F.F. 1979. The Dental History of the Manchester Mummies. *In:* A.R. DAVID. Ed. *The Manchester Museum Mummy Project.* Manchester Museum, Manchester. 65-79.

LEEK, F.F. 1984. Dental problems during the Old Kingdom. *In:* A.R. DAVID and E. TAPP, Eds., *Evidence Embalmed.* Manchester University Press. Manchester. 104-133.

LEEK, F.F. 1986. Dental health and disease in ancient Egypt with reference to the Manchester mummies. *In:* A.R. DAVID, Ed. *Science in Egyptology.* Manchester University Press. Manchester. 35-48.

LEVERS, B.G.H. and DARLING, A.J. 1983. Continuous eruption of some adult human teeth of ancient populations. *Archives of Oral Biology.* **28.** No.5. 401-408.

LISOWSKI, F.P. 1967. Prehistoric and early historic Trepanation. *In:* D. BROTHWELL AND A.T. SANDISON., Eds., *Diseases in Antiquity.* Charles Thomas, Springfield, Illinois, U.S.A., 651-673.

LODGE, T. 1967. Thinning of the parietal bones in Early Egyptian populations and its aetiology in the light of modern observations. *In:* D. BROTHWELL AND A.T. SANDISON., Eds., *Diseases in Antiquity.* Charles C. Thomas, Springfield, Illinois, U.S.A. 405-413.

LYNCH, M.A. 1977. *Burket's Oral Medicine.* 7th Edition. J.P.Lippencott Co., Philadelphia.

MAEGRAITH, B. 1989 Nutritional Disorders. *In:* ADAMS and MAEGRAITH, Eds., *Clinical Tropical Diseases.* 9th Edition. Blackwell Scientific Publications, 256-280.

MANN, G.E. 1989. On the accuracy of sexing of skeletons in archaeological reports. Brief Communications. *Journal of Egyptian Archaeology.* **75,** 246-249.

MANNICHE, L. 1993. *An Ancient Egyptian Herbal.* British Museum Press, London.

MAYS, S., RUA, C. de la and MOLLESON, T. 1995. Molar crown height as a means of evaluating existing dental wear scales for estimating age at death in human skeletal remains. *Journal of Archaeological Science.* **22,** 659-670.

MIDANT-REYNES, B. 1992. Introduction. *The Prehistory of Egypt.* Blackwell Publishers, Oxford. 1-12.

MILES, A.E.W. 1962. Assessment of the ages of a population of Anglo-Saxons from their dentitions. *Proceedings of the Royal society of Medicine.* **55,** 881-886.

MILLER, J. 1997. *Dislocation of the mandible in the ancient Egyptian and its possible aetiology.* Unpublished M.Sc. dissertation. Manchester University.

MILLER, J. and McDADE, C. 1992. The Dental Examination of Natsef Amun. *In:* A.R.. DAVID and E. TAPP, Eds. *The Mummy's Tale.* Michael O'Mara, London, 112-121.

MILLET, N.B., HART, G.D., REYMAN, T.A, ZIMMERMAN, A.R. and LEWIN, P.K. 1980. ROM 1: mummification for the common people. *In:* A. and E. COCKBURN, Eds., *Mummies, Disease and Ancient Cultures.* Cambridge University Press, Cambridge. 71-84.

MOFFETT, B.J., JOHNSON, L.C., McCABE, J.B. and ASKEW, H.C. 1964. Articular remodelling in the adult human temporomandibular joint. *American Journal of Anatomy,* **115**, 119-130.

MOLLESON, T. 2001. *Personal communication from the palaeopathology department at the Natural History Museum.*

MOLNAR, S. 1972. Human Tooth Wear, Tooth Function and cultural variability. *American Journal of Physical Anthropology,* **34,** 175-190.

MOWAT, A.P. 1987. *Non-viral infections of the liver.* Liver disorders in childhood. Butterworth, London.

MUIR, R.N.M. 1992. *Muir's Textbook of Pathology.* 13th Edition. Edward Arnold, London.

MURRAY, M. 1972. History. *The splendour that was Egypt.* Praeger Paperbacks, New York, 11-57.

MURPHY, B. 1998. *Personal Communication.*

MURPHY, T. 1959. The changing pattern of dentine exposure in human tooth attrition. *American Journal of Physical Anthropology.* **17,** 167-178.

NUNN, J.F. 1996. *Ancient Egyptian Medicine.* British Museum Press, London.

OBERG, T., CARLSSON, G.E., and FAJERS, C.M. 1971. The temporomandibular joint. A morphological study on human autopsy material. *Acta Odontologica Scandinavia.* **29,** 349-383.

ORTNER, D.J. 1984. Bone lesions in a probable case of scurvy from Metlatavik, Alaska. *Museum of applied science centre of archaeology.* **3,** 79-81.

OSMAN, A. 1989. *Stranger in the Valley of the Kings.* Paladin Grafton Books, Collins Publishing Group, London.

PORTER, R. 1997. The roots of medicine. *The Greatest Benefit to Mankind.* Harper Collins, London. 18-20.

QUIRKE, S. 1999. *Personal Communication.*

RAWLINGS, B. 1999. *Personal Communication.*

REISNER, G.A. 1905. *The Hearst Medical Papyrus.* J.C. Hinrichs, Leipzig.

ROHL, D. 1995. *A Test of Time.* Century Random House UK. Ltd., London. 333.

RICHARDS, L.C. and BROWN, T. 1981. Dental attrition and degenerative arthritis of the temporomandibular joint. *Journal of Oral Rehabilitation.* **8,** 293-307.

ROBERTS, C. 1987. Scurvy. *Palaeopathology Association Newsletter.* **57,** 14-15.

ROBERTS, C. and MANCHESTER, K. 1995. Back to Basics. *The Archaeology of Disease.* Alan Sutton Publishing Ltd., Ithaca, NY. 15-29.

ROBERTS, C. and MANCHESTER, K. 1995 [2]. Dental Disease. *The Archaeology of Disease.* Alan Sutton Publishing Ltd., Ithaca, NY. 44-64.

RUFFER, M.A. 1920. Studies of abnormalities and pathology of ancient Egyptian teeth. *American Journal of Physical Anthropology.* **3,** 335-382.

RUSSELL, W., STOREY, A.T. and PONITZ. 1980. Radiographic techniques in the study of the mummy. *An X-Ray Atlas of the Royal Mummies. In:* J.E. HARRIS and E.F. WENTE, Eds., University of Chicago Press, Chicago. 163-188.

SAFFIRIO, L. 1972. Food and Dietary Habits in Ancient Egypt. *Journal of Human Evolution.* **1,** 297-305.

SAMUEL, D. 1993. Ancient Egyptian Bread and Beer, An Interdisciplinary Approach. *In:* W.V. DAVIES and R. WALKER, Eds. *Biological Anthropology and the study of Ancient Egypt.* British Museum Press, London. 156-164.

SANDISON, A.T. 1967 Parasitic Diseases. *In:* D. BROTHWELL and A.T. SANDISON, Eds., *Diseases in Antiquity.* Charles C. Thomas, Springfield, Illinois, U.S.A., 178-184.

SATINOFF, M.L. 1972. The Medical Biology of the Early Egyptian Populations from Aswan, Assyut and Gebelen. *Journal of Human Evolution*, **1,** 247-257.

SCHOUR, I. and MASSLER, M. 1941. The Development of the Human Dentition. *Journal of the American Dental Association,* **28,** 1153-1160.

SCOTT, J.H. and SYMONS, N.B.B. 1977. A Summary of the Age Changes in the Teeth and the Jaws. 8th Edition. *Dental Anatomy.* Churchill Livingstone. Edinburgh, London and NY. 333-338.

SELINCOURT, A. de. 1972. *Herodotus. The Histories. Book 2.* Penguin Books. London.

SEWARD, F.E. 1976. Tooth attrition and the temporomandibular joint. *Angle Orthodontics.* **46,** 162-170.

SHAFER, W.G., HINE, M.K., and LEVY, B.M. 1958 Regressive Alterations of the Teeth. *Oral Pathology.* W.B.Saunders Co., Philadelphia, 206-230.

SHAFER, W.G., HINE, M.K., and LEVY, B.M. 1958[2]. Oral Aspects of Metabolic Disease. *Oral Pathology.* W.B.Saunders Co., Philadelphia, 477-538.

SIGERIST, H.E. 1951. *A History of Medicine.* Oxford University Press. Oxford.

SMITH, N.J.D. 1979. Dental Pathology in an Ancient Egyptian Population. *In*: A.R. DAVID, Ed., *Science in Egyptology.* Manchester University Press, Manchester. 43-49.

SNOW, CLYDE C. 1982. Forensic Anthropology. *In:* J.A. COTTONE and S.M. STANDISH, Eds. *Outline of Forensic Dentistry.* Year Book Medical Publishers, Inc. USA. 34-53.

SOAMES, J.V. and SOUTHAM, J.C. 1998. Oral Pathology. 3rd edition. Oxford University Press, Oxford.

SOUHAMI, R.L. and MOXHAM, J. 1990. *Textbook of Medicine.* Churchill and Livingstone. London.

STETTER, C. 1993. Introduction. *The Secret Medicine of the Pharaohs.* Edition q., Berlin. 7-20.

STEVENS, J.M. 1975. Gynaecology from Ancient Egypt: The Papyrus Kahun. *The Medical Journal of Australia.* **2,** 949-952.

STROUHAL, E. 1976. Tumours in the Remains of Ancient Egyptians. *American Journal of Physical Anthropology.* **45,** 613-620.

STROUHAL, E. 1985. Embalming Excerebration. *In:* A.R. DAVID, Ed., *Science in Egyptology.* Manchester University Press, Manchester. 141-154.

TAPP, E. 1979. Disease in the Manchester Mummies. *In:* A.R. DAVID, Ed., *The Manchester Mummy Project.* Manchester Museum, Manchester. 95-102.

TAPP, E. and WILDSMITH, K. 1992. The Autopsy and Endoscopy of the Leeds Mummy. *In:* A.R. DAVID and E. TAPP, Eds., *A Mummy's Tale.* Michael O'Mara Books Ltd., London. 132-154.

THOMPSON, C.E. 2001. *Hair and Henna in Ancient Egypt; a Mass Spectrometry Analysis.* Unpublished MSc. Thesis, Manchester University.

WALDRON, T. 2001. Infectious Diseases. *Shadows in the Soil.* Tempus Publishing, Stroud, Gloucestershire, 97-104.

WATSON, P. 1987. *Egyptian Pyramids and Mastaba Tombs.* Shire Publications Ltd., Princes Risborough, Buckinghamshire.

WEDEL, A., CARLSSON, G.E. and SAGNE, S. 1978. Temporomandibular joint morphology in a medieval skull material. *Swedish Dental Journal.* **2,** 177-187.

WEEKS, K.R. 1980. Ancient Egyptian Dentistry. *In:* J.E. HARRIS and E.F. WENTE, Eds., *An X-Ray Atlas of the Royal Mummies.* University of Chicago Press, Chicago. 99-121.

WEINBERGER, B.W. 1947. The Dental Art in Ancient Egypt. *Journal of the American Dental Association.* **34,** 170-184.

WEINBERGER, B.W. 1948. Ancient Medical Concept of Disease. *History of Dentistry.* C.V.Mosby Co., St. Louis. 22-25.

WELLS, C. 1967. Pseudopathology. *In:* D. BROTHWELL and A.T. SANDISON, Eds., *Diseases in Antiquity.* Charles C. Thomas, Springfield, Illinois, U.S.A., 8-20.

WENTE, E.F. 1995. Who was who among the Royal Mummies. *The Oriental Institute News and Notes, No.144.* (From internet www. oi.uchicago.edu/OI/IS/WENTE/NN_Win95.html)

WHITTAKER, D.K., PARKER, J.H. and JENKINS, C. 1982. Tooth attrition and continuing eruption in a Romano-British Population. *Archives of Oral Biology.* **27,** 405-409.

WHITTAKER, D.K., DAVIES, G., and BROWN, M. 1985. Tooth loss, attrition and temporomandibular joint changes in a Romano-British population. *Journal of Oral Rehabilitation.* **12,** 407-419.

WHITTAKER, D.K., MOLLESON, T., DANIEL, A.T., WILLIAMS, J.T., ROSE, P. and RESTEGHINI, R. 1985. *Quantitative assessment of tooth wear, alveolar-crest height and continuing eruption in a Romano-British population.* Archives of Oral Biology. **30,** No.6. (493-501).

WHITTAKER, D.K., GRIFFITHS, S., ROBSON, A., ROGER-DAVIES, P., THOMAS, G. and MOLLESON, T.. 1990. Continuing tooth eruption and alveolar crest height in an eighteenth century population from Spitalfields, East London. *Archives of Oral Biology,* **35,** No.2, 81-85.

WHITTAKER, D.K., JONES, J.W., EDWARDS, P.W. and MOLLESON, T. 1990b. Studies of the temporomandibular joints of an eighteenth-century London population. *Journal of Oral Rehabilitation,* **17,** 89-97.

WILSON and KORNMAN. 1996. Systemic Modifiers. *Fundamentals of Periodontics.* Quintessence Publishing Co. Inc., Illinois.

ZIAS, J. and NUMEROFF, K. 1987. Operative Dentistry in the second century BCE. *Journal of the American Dental Association,* **114,** 665-666.

www.ingramcontent.com/pod-product-compliance
Lightning Source LLC
LaVergne TN
LVHW071218210926
848482LV00015B/841
* 9 7 8 1 4 0 7 3 0 2 8 2 9 *